NANOTECHNOLOGY AND THE ENVIRONMENT

NANOTECHNOLOGY AND THE ENVIRONMENT

ROBERT V. NEUMANN
EDITOR

Nova Science Publishers, Inc.
New York

For permission to use material from this book please contact us:
Telephone 631-231-7269; Fax 631-231-8175
Web Site: http://www.novapublishers.com

NOTICE TO THE READER

The Publisher has taken reasonable care in the preparation of this book, but makes no expressed or implied warranty of any kind and assumes no responsibility for any errors or omissions. No liability is assumed for incidental or consequential damages in connection with or arising out of information contained in this book. The Publisher shall not be liable for any special, consequential, or exemplary damages resulting, in whole or in part, from the readers' use of, or reliance upon, this material. Any parts of this book based on government reports are so indicated and copyright is claimed for those parts to the extent applicable to compilations of such works.

Independent verification should be sought for any data, advice or recommendations contained in this book. In addition, no responsibility is assumed by the publisher for any injury and/or damage to persons or property arising from any methods, products, instructions, ideas or otherwise contained in this publication.

This publication is designed to provide accurate and authoritative information with regard to the subject matter covered herein. It is sold with the clear understanding that the Publisher is not engaged in rendering legal or any other professional services. If legal or any other expert assistance is required, the services of a competent person should be sought. FROM A DECLARATION OF PARTICIPANTS JOINTLY ADOPTED BY A COMMITTEE OF THE AMERICAN BAR ASSOCIATION AND A COMMITTEE OF PUBLISHERS.

LIBRARY OF CONGRESS CATALOGING-IN-PUBLICATION DATA

Nanotechnology and the environment / [edited by] Robert V. Neumann.
  p. cm.
Includes bibliographical references and index.
ISBN 978-1-60692-663-5 (hardcover)
1. Environmental engineering. 2. Nanotechnology. I. Neumann, Robert V.
TA170.N35 2009
620'.5--dc22
              2009028449

Published by Nova Science Publishers, Inc. ✦ *New York*

CONTENTS

PREFACE[*]

Nanotechnology has potential applications in many sectors of the American economy, including consumer products, health care, transportation, energy and agriculture. Since 2001, EPA has played a leading role in funding research and setting research directions to develop environmental applications for, and understand the potential human health and environmental implications of, nanotechnology. That research has already borne fruit, particularly in the use of nanomaterials for environmental clean-up and in beginning to understand the disposition of nanomaterials in biological systems. Some environmental applications using nanotechnology have progressed beyond the research stage. The purpose of this book is elucidate the needs associated with nanotechnology, to support related EPA programs, and to communicate these nanotechnology science issues to stakeholders and the public. The book begins with an introduction that describes what nanotechnology is, why EPA is interested in it, and what opportunities and challenges exist regarding nanotechnology and the environment. It then moves to a discussion of the potential environmental benefits of nanotechnology, describing environmental technologies as well as other applications that can foster sustainable use of resources. The book next provides an overview of existing information on nanomaterials regarding components needed to conduct a risk assessment.

---

[*] This is an edited, reformatted and augmented edition of a United States Environmental Protection Agency publication, EPA 100/B-07/001, dated February 2007.

In: Nanotechnology and the Environment
Editor: Robert V. Neumann

ISBN: 978-1-60692-663-5
© 2010 Nova Science Publishers, Inc.

NANOTECHNOLOGY WHITE PAPER

U.S. Environmental Protection Agency

FOREWORD

Nanotechnology presents opportunities to create new and better products. It also has the potential to improve assessment, management, and prevention of environmental risks. However, there are unanswered questions about the impacts of nanomaterials and nanoproducts on human health and the environment.

In December 2004, EPA's Science Policy Council (SPC) formed a cross-Agency Nanotechnology Workgroup to develop a white paper examining potential environmental applications and implications of nanotechnology. This document describes the issues that EPA should consider to ensure that society benefits from advances in environmental protection that nanotechnology may offer, and to understand and address any potential risks from environmental exposure to nanomaterials. Nanotechnology will have an impact across EPA. Agency managers and staff are working together to develop an approach to nanotechnology that is forward thinking and informs the risk assessment and risk management activities in our program and regional offices. This document is intended to support that cross-Agency effort.

We would like to acknowledge and thank the Nanotechnology Workgroup subgroup co¬chairs and members and for their work in developing this document. We would especially like to acknowledge the Workgroup co-chairs Jim Willis and Jeff Morris for leading the workgroup and document development. We also thank SPC staff task lead Kathryn Gallagher, as well as Jim Alwood, Dennis Utterback, and Jeremiah Duncan for their efforts in assembling and refining the document.

It is with pleasure that we provide the EPA Nanotechnology White Paper to promote the use of this new, exciting technology in a manner that protects human health and the environment.

William H. Benson
Acting Chief Scientist
Office of the Science Advisor

Charles M. Auer
Director, Office of Pollution
Prevention and Toxics

ACKNOWLEDGMENTS

The Nanotechnology Workgroup would like to acknowledge the Science Policy Council and its Steering Committee for their recommendations and contributions to this document. We thank Paul Leslie of TSI Incorporated, and Laura Morlacci, Tom Webb and Peter McClure of Syracuse Research Corporation for their support in developing background information for the document. We also thank the external peer reviewers (listed in an appendix) for their comments and suggestions. Finally, the workgroup would like to thank Bill Farland and Charles Auer for their leadership and vision with respect to nanotechnology.

EXECUTIVE SUMMARY

Nanotechnology has potential applications in many sectors of the American economy, including consumer products, health care, transportation, energy and agriculture. In addition, nanotechnology presents new opportunities to improve how we measure, monitor, manage, and minimize contaminants in the environment. While the U.S. Environmental Protection Agency (EPA, or "the Agency") is interested in researching and developing the possible benefits of nanotechnology, EPA also has the obligation and mandate to protect human health and safeguard the environment by better understanding and addressing potential risks from exposure to nanoscale materials and products containing nanoscale materials (both referred to here as "nanomaterials").

Since 2001, EPA has played a leading role in funding research and setting research directions to develop environmental applications for, and understand the potential human health and environmental implications of, nanotechnology. That research has already borne fruit, particularly in the use of nanomaterials for environmental clean-up and in beginning to understand the disposition of nanomaterials in biological systems. Some environmental applications using nanotechnology have progressed beyond the research stage. Also, a number of specific nanomaterials have come to the Agency's attention, whether as novel products intended to promote the reduction or remediation of pollution or because they have entered one of EPA's regulatory review processes. For EPA, nanotechnology has evolved from a futuristic idea to watch, to a current issue to address.

In December 2004, EPA's Science Policy Council created a cross-Agency workgroup charged with describing key science issues EPA should consider to ensure that society accrues the important benefits to environmental protection that nanotechnology may offer, as well as to better understand any potential risks from exposure to nanomaterials in the environment. This paper is the product of that workgroup.

The purpose of this paper is to inform EPA management of the science needs associated with nanotechnology, to support related EPA program office needs, and to communicate these nanotechnology science issues to stakeholders and the public. The paper begins with an introduction that describes what nanotechnology is, why EPA is interested in it, and what opportunities and challenges exist regarding nanotechnology and the environment. It then moves to a discussion of the potential environmental benefits of nanotechnology, describing environmental technologies as well as other applications that can foster sustainable use of resources. The paper next provides an overview of existing information on nanomaterials regarding components needed to conduct a risk assessment. Following that there is a brief section on responsible development and the Agency's statutory mandates. The paper then provides an extensive review of research needs for both environmental applications and implications of nanotechnology. To help EPA focus on priorities for the near term, the paper concludes with staff recommendations for addressing science issues and research needs, and includes prioritized research needs within most risk assessment topic areas (e.g., human health effects research, fate and transport research). In a separate follow-up effort to this White Paper, EPA's Nanotechnology Research Framework, attached in Appendix C of this paper, was developed by EPA's Office of Research and Development (ORD) Nanotechnology Research Strategy Team. This team is composed of representatives from across ORD. The Nanotechnology Research Framework outlines how EPA will strategically focus its own research program to provide key information on potential environmental impacts from human or ecological exposure to nanomaterials in a manner that complements other federal, academic, and private-sector research activities. Additional supplemental information is provided in a number of other appendices.

Key Nanotechnology White Paper recommendations include:

- **Environmental Applications Research.** The Agency should continue to undertake, collaborate on, and support research to better understand and apply information regarding environmental applications of nanomaterials.

- **Risk Assessment Research.** The Agency should continue to undertake, collaborate on, and support research to better understand and apply information regarding nanomaterials':

  - chemical and physical identification and characterization,
  - environmental fate,
  - environmental detection and analysis,
  - potential releases and human exposures,
  - human health effects assessment, and
  - aecological effects assessment.

To ensure that research best supports Agency decision making, EPA should conduct case studies to further identify unique risk assessment considerations for nanomaterials.

- **Pollution Prevention, Stewardship, and Sustainability.** The Agency should engage resources and expertise to encourage, support, and develop approaches that promote pollution prevention, sustainable resource use, and good product stewardship in the production, use and end of life management of nanomaterials. Additionally, the Agency should draw on new, "next generation" nanotechnologies to identify ways to support environmentally beneficial approaches such as green energy, green design, green chemistry, and green manufacturing.

- **Collaboration and Leadership.** The Agency should continue and expand its collaborations regarding nanomaterial applications and potential human health and environmental implications.

- **Intra-Agency Workgroup.** The Agency should convene a standing intra-Agency group to foster information sharing on nanotechnology science and policy issues.

- **Training.** The Agency should continue and expand its nanotechnology training activities for scientists and managers.

Nanotechnology has emerged as a growing and rapidly changing field. New generations of nanomaterials will evolve, and with them new and possibly unforeseen environmental issues. It will be crucial that the Agency's approaches to leveraging the benefits and assessing the impacts of nanomaterials continue to evolve in parallel with the expansion of and advances in these new technologies.

In: Nanotechnology and the Environment
Editor: Robert V. Neumann

ISBN: 978-1-60692-663-5
© 2010 Nova Science Publishers, Inc.

Chapter 1

INTRODUCTION

U.S. Environmental Protection Agency

1.1. PURPOSE

Nanotechnology presents potential opportunities to create better materials and products. Already, nanomaterial-containing products are available in U.S. markets including coatings, computers, clothing, cosmetics, sports equipment and medical devices. A survey by EmTech Research of companies working in the field of nanotechnology has identified approximately 80 consumer products, and over 600 raw materials, intermediate components and industrial equipment items that are used by manufacturers (Small Times Media, 2005). A second survey by the Project on Emerging Nanotechnologies at the Woodrow Wilson International Center for Scholars lists over 300 consumer products (http://www.nanote chproject.org/index.php?id=44 or http://www.nanotechproject.org/consumerproducts). Our economy will be increasingly affected by nanotechnology as more products containing nanomaterials move from research and development into production and commerce.

Nanotechnology also has the potential to improve the environment, both through direct applications of nanomaterials to detect, prevent, and remove pollutants, as well as indirectly by using nanotechnology to design cleaner industrial processes and create environmentally responsible products. However, there are unanswered questions about the impacts of nanomaterials and nanoproducts on human health and the environment, and the U.S. Environmental Protection Agency (EPA or "the Agency") has the obligation to ensure that potential risks are adequately understood to protect human health and the environment. As products made from nanomaterials become more numerous and therefore more prevalent in the environment, EPA is thus considering how to best leverage advances in nanotechnology to enhance environmental protection, as well as how the introduction of nanomaterials into the environment will impact the Agency's environmental programs, policies, research needs, and approaches to decision making.

In December 2004, the Agency's Science Policy Council convened an intra-Agency Nanotechnology Workgroup and charged the group with developing a white paper to examine the implications and applications of nanotechnology. This document describes key science issues EPA should consider to ensure that society accrues the benefits to environmental protection that nanotechnology may offer and that the Agency understands and addresses potential risks from environmental exposure to nanomaterials.

The purpose of this paper is to inform EPA management of the science needs associated with nanotechnology, to support related EPA program office needs, and to communicate these nanotechnology science issues to stakeholders and the public. The paper begins with an introduction that describes what nanotechnology is, why EPA is interested in it, and what opportunities and challenges exist regarding nanotechnology and the environment. It then moves to a discussion of the potential environmental benefits of nanotechnology, describing environmental technologies as well as other applications that can foster sustainable use of resources. The paper next provides an overview of existing information on nanomaterials regarding components needed to conduct a risk assessment. Following that is a brief section on responsible development and the Agency's statutory mandates. The paper then provides an extensive review of research needs for both environmental applications and implications of nanotechnology. To help EPA focus on priorities for the near term, the paper concludes with staff recommendations for addressing science issues and research needs, and includes prioritized research needs within most risk assessment topic areas (e.g., human health effects research, fate and transport research). In a separate follow-up effort to this White Paper, EPA's Nanotechnology Research Framework, attached in Appendix C of this paper, was developed by EPA's Office of Research and Development (ORD) Nanotechnology Research Strategy Team. This team is composed of representatives from across ORD. The Nanotechnology Research Framework outlines how EPA will strategically focus its own research program to provide key information on potential environmental impacts from human or ecological exposure to nanomaterials in a manner that complements other federal, academic, and private-sector research activities. Additional supplemental information is provided in a number of additional appendices.

A discussion of an entire technological process or series of processes, as is nanotechnology, could be wide ranging. However, because EPA operates through specific programmatic activities and mandates, this document confines its discussion of nanotechnology science issues within the bounds of EPA's statutory responsibilities and authorities. In particular, the paper discusses what scientific information EPA will need to address nanotechnology in environmental decision making.

1.2. NANOTECHNOLOGY DEFINED

A nanometer is one billionth of a meter (10^{-9} m)—about one hundred thousand times smaller than the diameter of a human hair, a thousand times smaller than a red blood cell, or about half the size of the diameter of DNA. Figure 1 illustrates the scale of objects in the nanometer range. For the purpose of this document, nanotechnology is defined as: research and technology development at the atomic, molecular, or macromolecular levels using a length scale of approximately one to one hundred nanometers in any dimension; the creation and use of structures, devices and systems that have novel properties and functions because of their small size; and the ability to control or manipulate matter on an atomic scale. This definition is based on part on the definition of nanotechnology used by the National

Nanotechnology Initiative (NNI), a U.S. government initiative launched in 2001 to coordinate nanotechnology research and development across the federal government (NNI, 2006a, b, c).

Nanotechnology is the manipulation of matter for use in particular applications through certain chemical and / or physical processes to create materials with specific properties. There are both "bottom-up" processes (such as self-assembly) that create nanoscale materials from atoms and molecules, as well as "top-down" processes (such as milling) that create nanoscale materials from their macro-scale counterparts. Figure 2 shows an example of a nanomaterial assembled through "bottom-up" processes. Nanoscale materials that have macro-scale counterparts frequently display different or enhanced properties compared to the macro-scale form. For the remainder of this document such engineered or manufactured nanomaterials will be referred to as "intentionally produced nanomaterials," or simply "nanomaterials." The definition of nanotechnology does not include unintentionally produced nanomaterials, such as diesel exhaust particles or other friction or airborne combustion byproducts, or nanosized materials that occur naturally in the environment, such as viruses or volcanic ash. Where information from incidentally formed or natural nanosized materials (such as ultrafine particulate matter) may aid in the understanding of intentionally produced nanomaterials, this information will be discussed, but the focus of this document is on intentionally produced nanomaterials.

There are many types of intentionally produced nanomaterials, and a variety of others are expected to appear in the future. For the purpose of this document, most current nanomaterials could be organized into four types:

(1) **Carbon-based materials.** These nanomaterials are composed mostly of carbon, most commonly taking the form of a hollow spheres, ellipsoids, or tubes. Spherical and ellipsoidal carbon nanomaterials are referred to as fullerenes, while cylindrical ones are called nanotubes. These particles have many potential applications, including improved films and coatings, stronger and lighter materials, and applications in electronics. Figures 3, 4, and 5 show examples of carbon-based nanomaterials.

(2) **Metal-based materials.** These nanomaterials include quantum dots, nanogold, nanosilver and metal oxides, such as titanium dioxide. A quantum dot is a closely packed semiconductor crystal comprised of hundreds or thousands of atoms, and whose size is on the order of a few nanometers to a few hundred nanometers. Changing the size of quantum dots changes their optical properties. Figures 6 and 7 show examples of metal-based nanomaterials.

(3) **Dendrimers.** These nanomaterials are nanosized polymers built from branched units. The surface of a dendrimer has numerous chain ends, which can be tailored to perform specific chemical functions. This property could also be useful for catalysis. Also, because three-dimensional dendrimers contain interior cavities into which other molecules could be placed, they may be useful for drug delivery. Figure 8 shows an example a dendrimer.

(4) **Composites** combine nanoparticles with other nanoparticles or with larger, bulk-type materials. Nanoparticles, such as nanosized clays, are already being added to products ranging from auto parts to packaging materials, to enhance mechanical, thermal, barrier, and flame-retardant properties. Figure 9 shows an example of a composite.

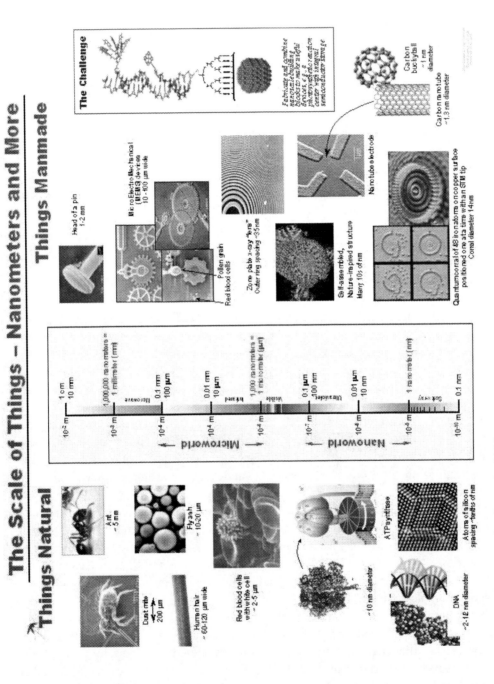

Figure 1. Diagram indicating relative scale of nanosized objects.
(From NNI website, courtesy Office of Basic Energy Sciences, U.S. Department of Energy.)

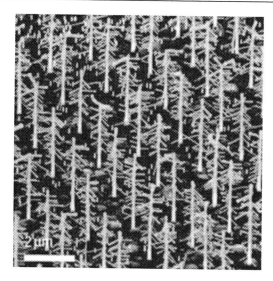

Figure 2. Gallium Phosphide (GaP) Nanotrees.
Semiconductor nanowires produced by controlled seeding, vapor-liquid-solid self-assembly. Bottom-up processes used for control over size and morphology. (Image used by permission, Prof. Lars Samuelson, Lund University, Sweden. [Dick et al. 2004])

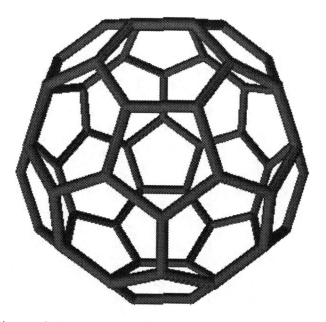

Figure 3. Computer image of aC-60 Fullerene. U.S. EPA.

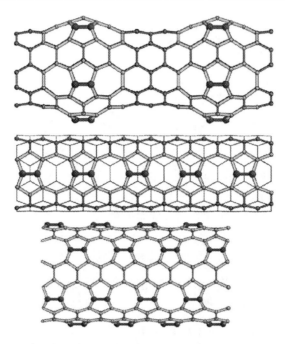

Figure 4. Computer images of various forms of carbon nanotubes.
(Images courtesy of Center for Nanoscale Materials, (Images courtesy of Center for Nanoscale Materials, Argonne National Laboratory)

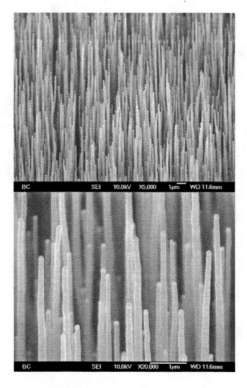

Figure 5. "Forest" of aligned carbon nanotubes.
(Image courtesy David Carnahan of NanoLab, Inc.)

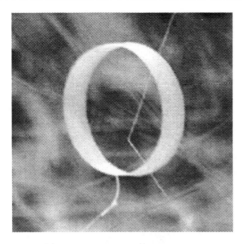

Figure 6. Zinc oxide nanostructure synthesized by a vapor-solid process.
(Image courtesy of Prof. Zhong Lin Wang, Georgia Tech)

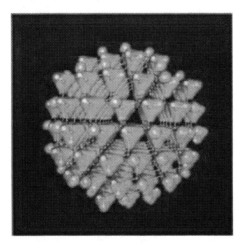

Figure 7. Computer image of a Gallium arsenide quantum dot of 465 atoms.
(Image courtesy of Lin-Wang Wang, Lawrence Berkeley National Laboratory)

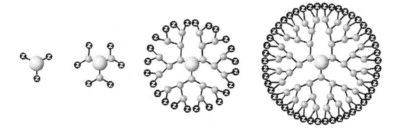

Figure 8. Computer image of generations of a dendrimer.
Dendrimers are nanoscale branched polymers that are grown in a stepwise fashion, which allows for precise control of their size. (Image courtesy of Dendritic NanoTechnologies, Inc.)

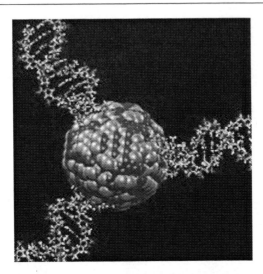

Figure 9. Computer image of a nano-biocomposite.
Image of a titanium molecule (center) with DNA strands attached, a bio-inorganic composite. This kind of material has potential for new technologies to treat disease. (Image courtesy of Center for Nanoscale Materials, Argonne National Lab)

The unique properties of these various types of intentionally produced nanomaterials give them novel electrical, catalytic, magnetic, mechanical, thermal, or imaging features that are highly desirable for applications in commercial, medical, military, and environmental sectors. These materials may also find their way into more complex nanostructures and systems as described in Figure 10. As new uses for materials with these special properties are identified, the number of products containing such nanomaterials and their possible applications continues to grow. Table 1 lists some examples of nanotechnology products listed in the Woodrow Wilson Center Consumer Products Inventory (http://www.nanotechproject.org/ 44/consumer-nanotechnology). There are estimates that global sales of nanomaterials could exceed $1 trillion by 2015 (M.C. Roco, presentation to the National Research Council, 23 March 2005, presentation available at http://www.nsf.gov/crssprgm/nano/reports/nnipres.jsp).

1.2.1 Converging Technologies

In the long-term, nanotechnology will likely be increasingly discussed within the context of the convergence, integration, and synergy of nanotechnology, biotechnology, information technology, and cognitive technology. Convergence involves the development of novel products with enhanced capabilities that incorporate bottom-up assembly of miniature components with accompanying biological, computational and cognitive capabilities. The convergence of nanotechnology and biotechnology, already rapidly progressing, will result in the production of novel nanoscale materials. The convergence of nanotechnology and biotechnology with information technology and cognitive science is expected to rapidly accelerate in the coming decades. The increased understanding of biological systems will provide valuable information towards the development of efficient and versatile biomimetic tools, systems, and architecture.

Table 1. Examples of Products that Use Nanotechnology and Nanomaterials

Health and Fitness	Electronics and Computers	Home and Garden	Food and Beverage	Other
Wound dressing	Computer displays	Paint	Non-stick coatings for pans	Coatings
Pregnancy test	Games	Antimicrobial pillows		Lubricants
Toothpaste			Antimicrobial refrigerator	
Golf club	Computer hardware	Stain resistant cushions	Canola oil	
Tennis Racket				
Skis				
Antibacterial socks				
Waste and stain resistant pants				
Cosmetics				
Air filter				
Sunscreen				

Source: Woodrow Wilson Center Consumer Products Inventory.
(http://www.nanotechproject.org/44/consumer-nanotechnology)

Generally, biotechnology involves the use of microorganisms, or bacterial factories, which contain inherent "blueprints" encoded in the DNA, and a manufacturing process to produce molecules such as amino acids. Within these bacterial factories, the organization and self-assembly of complex molecules occurs routinely. Many "finished" complex cellular products are < 100 nanometers. For this reason, bacterial factories may serve as models for the organization, assembly and transformation for other nanoscale materials production.

Bacterial factory blueprints are also flexible. They can be modified to produce novel nanobiotechnology products that have specific desired physical-chemical (performance) characteristics. Using this production method could be a more material and energy efficient way to make new and existing products, in addition to using more benign starting materials. In this way, the convergence of nano- and biotechnologies could improve environmental protection. As an example, researchers have extracted photosynthetic proteins from spinach chloroplasts and coated them with nanofilms that convert sunlight to electrical current, which one day may lead to energy generating films and coatings (Das et al., 2004). The addition of information and cognitive capabilities will provide additional features including programmability, miniaturization, increased power capacities, adaptability, and reactive, self-correcting capacities.

Another example of converging technologies is the development of nanometer-sized biological sensor devices that can detect specific compounds within the natural environment; store, tabulate, and process the accumulated data; and determine the import of the data, providing a specific response for the resolved conditions would enable rapid and effective human health and environmental protection. Responses could range from the release of a certain amount of biological or chemical compound, to the removal or transformation of a compound.

The convergence of nanotechnology with biotechnology and with information and cognitive technologies may provide such dramatically different technology products that the manufacture, use and recycling/disposal of these novel products, as well as the development of policies and regulations to protect human health and the environment, may prove to be a daunting task.

The Agency is committed to keeping abreast of emerging issues within the environmental arena, and continues to support critical research, formulate new policies, and adapt existing policies as needed to achieve its mission. However, the convergence of these technologies will demand a flexible, rapid and highly adaptable approach within EPA. As these technologies progress and as novel products emerge, increasingly the Agency will find that meeting constantly changing demands depends on taking proactive actions and planning.

We may be nearing the end of basic research and development on the first generation of materials resulting from nanotechnologies that include coatings, polymers, more reactive catalysts, etc. (Figure 10). The second generation, which we are beginning to enter, involves targeted drug delivery systems, adaptive structures and actuators, and has already provided some interesting examples. The third generation, anticipated within the next 10-15 years, is predicted to bring novel robotic devices, three-dimensional networks and guided assemblies. The fourth stage is predicted to result in molecule-by-molecule design and self-assembly capabilities. Although it is not likely to happen for some time, this integration of these fourth-generation nanotechnologies with information, biological, and cognitive technologies will lead to products which can now only be imagined. While the Agency will not be able to predict the future, it needs to prepare for it. Towards that aim, understanding the unique challenges and opportunities afforded by converging technologies before they occur will provide the Agency with the essential tools for the effective and appropriate response to emerging technology and science.

1.3. WHY NANOTECHNOLOGY IS IMPORTANT TO EPA

Nanotechnology holds great promise for creating new materials with enhanced properties and attributes. These properties, such as greater catalytic efficiency, increased electrical conductivity, and improved hardness and strength, are a result of nanomaterials' larger surface area per unit of volume and quantum effects that occur at the nanometer scale ("nanoscale"). Nanomaterials are already being used or tested in a wide range of products such as sunscreens, composites, medical and electronic devices, and chemical catalysts. Similar to nanotechnology's success in consumer products and other sectors, nanomaterials have promising environmental applications. For example, nanosized cerium oxide has been developed to decrease diesel emissions, and iron nanoparticles can remove contaminants from soil and ground water. Nanosized sensors hold promise for improved detection and tracking of contaminants. In these and other ways, nanotechnology presents an opportunity to improve how we measure, monitor, manage, and reduce contaminants in the environment.

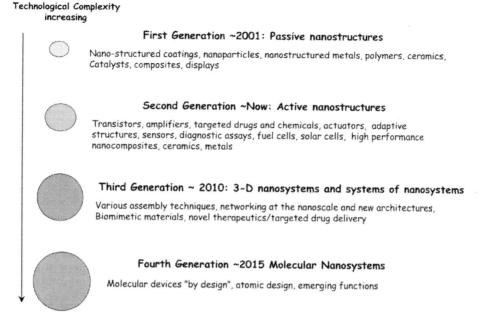

Figure 10. Projected Stages of Nanotechnology Development.
This analyis of the projected stages of nanotechnology development was first conceptualized by M.C. Roco.

Some of the same special properties that make nanomaterials useful are also properties that may cause some nanomaterials to pose hazards to humans and the environment, under specific conditions. Some nanomaterials that enter animal tissues may be able to pass through cell membranes or cross the blood-brain barrier. This may be a beneficial characteristic for such uses as targeted drug delivery and other disease treatments, but could result in unintended impacts in other uses or applications. Inhaled nanoparticles may become lodged in the lung or be translocated, and the high durability and reactivity of some nanomaterials raise issues of their fate in the environment. It may be that in most cases nanomaterials will not be of human health or ecological concern. However, at this point not enough information exists to assess environmental exposure for most engineered nanomaterials. This information is important because EPA will need a sound scientific basis for assessing and managing any unforeseen future impacts resulting from the introduction of nanoparticles and nanomaterials into the environment.

A challenge for environmental protection is to help fully realize the societal benefits of nanotechnology while identifying and minimizing any adverse impacts to humans or ecosystems from exposure to nanomaterials. In addition, we need to understand how to best apply nanotechnology for pollution prevention in current manufacturing processes and in the manufacture of new nanomaterials and nanoproducts, as well as in environmental detection, monitoring, and clean-up. This understanding will come from scientific information generated by environmental research and development activities within government agencies, academia, and the private sector.

1.4. NATIONAL AND INTERNATIONAL CONTEXT

EPA's role in nanotechnology exists within a range of activities by federal agencies and other groups that have been ongoing for several years. Figure 11 lists examples of federal sources of information and interaction to inform EPA's nanotechnology activities. Many sectors, including U.S. federal and state agencies, academia, the private-sector, other national governments, and international bodies, are considering potential environmental applications and implications of nanotechnology. This section describes some of the major players in this arena, for two principal reasons: to describe EPA's role regarding nanotechnology and the environment, and to identify opportunities for collaborative and complementary efforts.

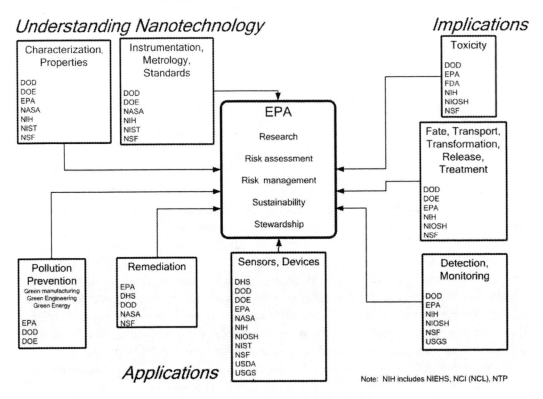

Figure 11. Federal Sources to Inform EPA's Nanotechnology Activities.
(Based on information in the NNI Supplement to the 2006 and 2007 budget and other information.)

1.4.1. Federal Agencies – The National Nanotechnology Initiative

The National Nanotechnology Initiative (NNI) was launched in 2001 to coordinate nanotechnology research and development across the federal government. Investments into federally funded nanotechnology-related activities, coordinated through the NNI, have grown from $464 million in 2001 to approximately $1.3 billion in 2006.

The NNI supports a broad range of research and development including fundamental research on the unique phenomena and processes that occur at the nano scale, the design and discovery of new nanoscale materials, and the development of nanotechnology-based devices and systems. The NNI also supports research on instrumentation, metrology, standards, and nanoscale manufacturing. Most important to EPA, the NNI has made responsible development of this new technology a priority by supporting research on environmental health and safety implications.

Twenty-five federal agencies currently participate in the NNI, thirteen of which have budgets which include to nanotechnology research and development. The other twelve agencies have made nanotechnology relevant to their missions or regulatory roles. Only a small part of this federal investment aims at researching the social and environmental implications of nanotechnology including its effects on human health, the environment, and society. Nine federal agencies are investing in implications research including the National Science Foundation, the National Institutes of Health, the National Institute for Occupational Health and Safety, and the Environmental Protection Agency. These agencies coordinate their efforts through the NNI's Nanoscale Science, Engineering, and Technology Subcommittee (NSET) and its Nanotechnology Environmental Health Implications workgroup (NEHI) (Figure 12). The President's Council of Advisors on Science and Technology (PCAST) has been designated as the national Nanotechnology Advisory Panel called for by the 21st Century Nanotechnology Research and Development Act of 2003. As such, PCAST is responsible for assessing and making recommendations for improving the NNI, including its activities to address environmental and other societal implications. The National Research Council also provides assessments and advice to the NNI.

Work under the NNI can be monitored through the website http://www.nano.gov.

1.4.2. Efforts of Other Stakeholders

About $2 billion in annual research and development investment are being spent by non-federal U.S. sectors such as states, academia, and private industry. State governments collectively spent an estimated $400 million on facilities and research aimed at the development of local nanotechnology industries in 2004 (Lux Research, 2004).

Although the industry is relatively new, the private sector is leading a number of initiatives. Several U.S. nanotechnology trade associations have emerged, including the NanoBusiness Alliance. The American Chemistry Council also has a committee devoted to nanotechnology and is encouraging research into the environmental health and safety of nanomaterials. In addition, the Nanoparticle Occupational Safety and Health Consortium has been formed by industry to investigate occupational safety and health issues associated with aerosol nanoparticles and workplace exposure monitoring and protocols. A directory of nanotechnology industry-related organizations can be found at http://www.nanovip.com.

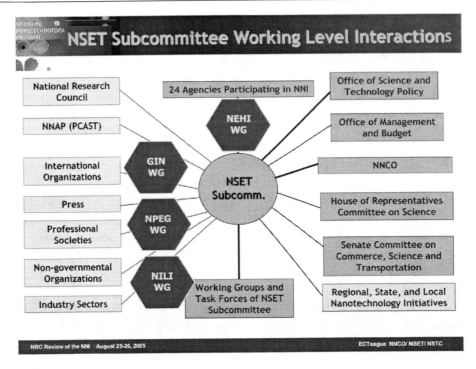

Figure 12. NNI NSET Subcommittee Structure

Environmental nongovernmental organizations (NGOs) such as Environmental Defense, Greenpeace UK, ETC Group, and the Natural Resources Defense Council are engaged in nanotechnology issues. Also, scientific organizations such as the National Academy of Sciences, the Royal Society of the United Kingdom, and the International Life Sciences Institute are providing important advice on issues related to nanotechnology and the environment.

1.4.3. International Activities

Fully understanding the environmental applications and implications of nanotechnology will depend on the concerted efforts of scientists and policy makers across the globe. Europe and Asia match or exceed the U.S. federal nanotechnology research budget. Globally, nanotechnology research and development spending is estimated at around $9 billion (Lux Research, 2006). Thus, a great opportunity exists for internationally coordinated and integrated efforts toward environmental research. Other governments have also undertaken efforts to identify research needs for nanomaterials (United Kingdom (UK) Department for Environment, Food and Rural Affairs, 2005; European Union Scientific Committee on Emerging and Newly Identified Health Risks (EU SCENIHR), 2005). International organizations such as the International Standards Organization and the Organisation for Economic Co-operation and Development (OECD) are engaged in nanotechnology issues. ISO has established a technical committee to develop international standards for nanotechnologies. This technical committee, ISO/TC 229 will develop standards for terminology and nomenclature, metrology and instrumentation, including specifications for reference materials, test methodologies, modeling and simulation, and science-based health, safety and environmental practices.

The OECD has engaged the topic of the implications of manufactured nanomaterials among its members under the auspices of the Joint Meeting of the Chemicals Committee and Working Party on Chemicals, Pesticides and Biotechnology (Chemicals Committee). On the basis of an international workshop hosted by EPA in Washington in December 2005, the Joint Meeting has agreed to establish a subsidiary body to work on the environmental health and safety implications of manufactured nanomaterials, with an eye towards enhancing international harmonization and burden sharing. In a related activity, the OECD's Committee on Scientific and Technology Policy is considering establishing a subsidiary body to address other issues related to realizing commercial and public benefits of advances in nanotechnology.

Additionally, the United States and European Union Initiative to Enhance Transatlantic Economic Integration and Growth (June 2005) addresses nanotechnology. Specifically, the Initiative states that the United States and the European Union will work together to, among other things, "support an international dialogue and cooperative activities for the responsible development and use of the emerging field of nanotechnology." EPA is also currently working with the U.S. State Department, the NNI, and the EU to bring about research partnerships in nanotechnology. Furthermore, in the context of environmental science, the EPA has worked with foreign research institutes and agencies (e.g., UK and Taiwan) to help inform nanotechnology and related environmental research programs.

By continuing to actively participate in international scientific fora, EPA will be well positioned to inform both domestic and international environmental policy. This will provide essential support for U.S. policy makers who work to negotiate international treaties and trade regimes. As products made from nanomaterials become more common in domestic and international channels of trade, policy makers will then be able to rely on EPA for the high quality science necessary to make effective decisions that could have a significant impact, both domestically and internationally, on human and environmental health, and economic well-being.

1.5. WHAT EPA IS DOING WITH RESPECT TO NANOTECHNOLOGY

EPA is actively participating in nanotechnology development and evaluation. Some of the activities EPA has undertaken include: 1) actively participating in the National Nanotechnology Initiative, which coordinates nanotechnology research and development across the federal government, 2) collaborating with scientists internationally in order to share the growing body of information on nanotechnology, 3) funding nanotechnology research through EPA's Science To Achieve Results (STAR) grant program and Small Business Innovative Research (SBIR) program and performing in-house research in the Office of Research and Development, 4) conducting regional nanotechnology research for remediation, 5) initiating the development of a voluntary program for the evaluation of nanomaterials and reviewing anomaterial premanufacture notifications in the Office of Pollution Prevention and Toxics, 6) reviewing nanomaterial registration applications in the Office of Air and Radiation/Office of Transportation and Air Quality, 7) reviewing potential nanoscale pesticides in the Office of Pesticide Programs, 8) investigating the use of nanoscale materials for environmental emediation in the Office of Solid Waste and Emergency Response; and 9) reviewing information and analyzing issues on nanotechnology in the Office of Enforcement and Compliance Assurance.

1.5.1. EPA's Nanotechnology Research Activities

Since 2001, EPA's ORD STAR grants program has funded 36 research grants nearly 12 million in the applications of nanotechnology to protect the environment, including the development of: 1) low-cost, rapid, and simplified methods of removing toxic contaminants from water, 2) new sensors that are more sensitive for measuring pollutants, 3) green manufacturing of nanomaterials; and 4) more efficient, selective catalysts. Additional applications projects have been funded through the SBIR program.

In addition, 14 recent STAR program projects focus on studying the possible harmful effects, or implications, of engineered nanomaterials. EPA has awarded or selected 30 grants to date in this area, totaling approximately $10 million. The most-recent research solicitations include partnerships with the National Science Foundation, the National Institute for Occupational Safety and Health, and the National Institute of Environmental Health Sciences. Research areas of interest for this proposal include the toxicology, fate, release and treatment, transport and transformation, bioavailability, human exposure, and life cycle assessment of nanomaterials. Appendix D lists STAR grants funded through 2005.

EPA's own scientists have done research in areas related to nanotechnology, such as on the toxicity of ultrafine particulate matter (e.g., Dreher, 2004). In addition, EPA scientists have begun to gather information on various environmental applications currently under development. ORD has also led development of an Agency Nanotechnology Research Framework for conducting and coordinating intramural and extramural nanotechnology research (Appendix C).

1.5.2. Regional Nanotechnology Research Activities for Remediation

A pilot study is planned at an EPA Region 5 National Priorities List site in Ohio. The pilot study will inject zero-valent iron nanoparticles into the groundwater to test its effectiveness in remediating volatile organic compounds. The study includes smaller pre-pilot studies and an investigation of the ecological effects of the treatment method. Information on the pilot can be found at http://www.epa.gov/region5/sites/nease/index.htm. Other EPA Regions (2, 3, 4, 9, and 10) are also considering the use of zero-valent iron in site remediation.

1.5.3. Office of Pollution Prevention and Toxics Activities Related to Nanoscale Materials

EPA's Office of Pollution Prevention and Toxics (OPPT) convened a public meeting in June 2005 regarding a potential voluntary pilot program for nanoscale materials. ("Nanoscale

Materials; Notice of Public Meeting," 70 Fed. Reg. 24574, May 10, 2005). At the meeting EPA received comment from a broad spectrum of stakeholders concerning all aspects of a possible stewardship program. Subsequently, OPPT invited the National Pollution Prevention and Toxics Advisory Committee (NPPTAC) to provide its views. NPPTAC established an Interim Ad Hoc Work Group on Nanoscale Materials which met in public to further discuss and receive additional public input on issues pertaining to the voluntary pilot program for nanoscale materials. The Interim Ad Hoc Work Group on Nanoscale Materials developed an overview document describing possible general parameters of a voluntary pilot program, which EPA is considering as it moves forward to develop and implement such a program. OPPT is already reviewing premanufacture notifications for a number of nanomaterials that have been received under the Toxics Substances Control Act (TSCA).

1.5.4. Office of Air and Radiation/Office of Transportation and Air Quality - Nanomaterials Registration Applications

EPA's Office of Air and Radiation/Office of Transportation and Air Quality has received and is reviewing an application for registration of a diesel additive containing cerium oxide. Cerium oxide nanoparticles are being marketed in Europe as on- and off-road diesel fuel additives to decrease emissions and some manufacturers are claiming fuel economy benefits.

1.5.5. Office of Pesticide Programs to Regulate Nano-Pesticide Products

Recently, members of the pesticide industry have engaged the Office of Pesticide Programs (OPP) regarding licensing/registration requirements for pesticide products that make use of nanotechnology. In response to the rapid emergence of these products, OPP is forming a largely intra-office workgroup to consider potential exposure and risks to human health and the ecological environment that might be associated with the use of nano-pesticides. Specifically, the workgroup will consider whether or not existing data are sufficient to support licensing/registration or if the unique characteristics associated with nano-pesticides warrant additional yet undefined testing. The workgroup will consider the exposure and hazard profiles associated with these new nano-pesticides on a case-by-case basis and ensure consistent review and regulation across the program.

1.5.6. Office of Solid Waste and Emergency Response

The Office of Solid Waste and Emergency Response (OSWER) is investigating potential implications and applications of nanotechnology that may affect its programs. In October 2005, OSWER worked with EPA's ORD and several other federal agencies to organize a Workshop on Nanotechnology for Site Remediation. The meeting summary and presentations from that workshop are available at http://www.frtr.gov/nano. In July 2006, OSWER held a symposium entitled, "Nanotechnology and OSWER: New Opportunities and Challenges." The symposium featured national and international experts, researchers, and industry leaders who discussed issues relevant to nanotechnology and waste management practices and focused on the life cycle of nanotechnology products. Information on the symposium will be posted on OSWER's website. OSWER's Technology Innovation and Field Services Division (TIFSD) is compiling a database of information on hazardous waste sites where project managers are considering using nanoscale zero-valent iron to address groundwater contamination. TIFSD is also preparing a fact sheet on the use of nanotechnology for site remediation that will be useful for site project managers. In addition, TIFSD has a website with links to relevant information on nanotechnology (http://clu-in.org/nano).

1.5.7. Office of Enforcement and Compliance Assurance

The Office of Enforcement and Compliance Assurance (OECA) is reviewing Agency information on nanotechnology (e.g., studies, research); evaluating existing statutory and regulatory frameworks to determine the enforcement issues associated with nanotechnology; evaluating the science issues for regulation/enforcement that are associated with nanotechnology, and; considering what information OECA's National Enforcement Investigations Center (NEIC) may need to consider to support the Agency.

1.5.7. Communication and Outreach

Gaining and maintaining public trust and support is important to fully realize the societal benefits and clearly communicate the impacts of nanotechnology. Responsible development of nanotechnology should involve and encourage an open dialogue with all concerned parties about potential risks and benefits. EPA is committed to keeping the public informed of the potential environmental impacts associated with nanomaterial development and applications. As an initial step, EPA is developing a dedicated web site to provide comprehensive information and enable transparent dialogue concerning nanotechnology. In addition, EPA has been conducting outreach by organizing and sponsoring sessions at professional society meetings, speaking at industry, state, and international nanotechnology meetings.

EPA already has taken steps to obtain public feedback on issues, alternative approaches, and decisions. For example, the previously noted OPPT public meetings were designed to discuss and receive public input. EPA will continue to work collaboratively with all stakeholders, including industry, other governmental entities, public interest groups, and the general public, to identify and assess potential environmental hazards and exposures resulting from nanotechnology, and to discuss EPA's roles in addressing issues of concern. EPA's goal is to earn and retain the public's trust by providing information that is objective, balanced, accurate and timely in its presentation, and by using transparent public involvement processes.

1.6. OPPORTUNITIES AND CHALLENGES

For EPA, the rapid development of nanotechnology and the increasing production of nanomaterials and nanoproducts present both opportunities and challenges. Using nanomaterials in applications that advance green chemistry and engineering and lead to the development of new environmental sensors and remediation technologies may provide us with new tools for preventing, identifying, and solving environmental problems. In addition, at this early juncture in nanotechnology's development, we have the opportunity to develop approaches that will allow us to produce, use, recycle, and eventually dispose of nanomaterials in ways that protect human health and safeguard the natural environment. The integration and synergy of nanotechnology, biotechnology, information technology, and cognitive technology will present opportunities as well as challenges to EPA and other regulatory agencies. To take advantage of these opportunities and to meet the challenge of ensuring the environmentally safe and sustainable development of nanotechnology, EPA must understand both the potential benefits and the potential impacts of nanomaterials and nanoproducts. The following chapters of this document discuss the science issues surrounding how EPA will gain and apply such understanding.

In: Nanotechnology and the Environment
Editor: Robert V. Neumann

ISBN: 978-1-60692-663-5
© 2010 Nova Science Publishers, Inc.

Chapter 2

ENVIRONMENTAL BENEFITS OF NANOTECHNOLOGY

U.S. Environmental Protection Agency

2.1. INTRODUCTION

As applications of nanotechnology develop over time, they have the potential to help shrink the human footprint on the environment. This is important, because over the next 50 years the world's population is expected to grow 50%, global economic activity is expected to grow 500%, and global energy and materials use is expected to grow 300% (World Resources Institute, 2000). So far, increased levels of production and consumption have offset our gains in cleaner and more-efficient technologies. This has been true for municipal waste generation, as well as for environmental impacts associated with vehicle travel, groundwater pollution, and agricultural runoff (OECD, 2001). This chapter will describe how nanotechnology can create materials and products that will not only directly advance our ability to detect, monitor, and clean-up environmental contaminants, but also help us avoid creating pollution in the first place. By more effectively using materials and energy throughout a product lifecycle, nanotechnology may contribute to reducing pollution or energy intensity per unit of economic output, reducing the "volume effect" described by the OECD.

2.2. BENEFITS THROUGH ENVIRONMENTAL TECHNOLOGY APPLICATIONS

2.2.1. Remediation/Treatment

Environmental remediation includes the degradation, sequestration, or other related approaches that result in reduced risks to human and environmental receptors posed by chemical and radiological contaminants such as those found at Comprehensive Environmental Response, Compensation and Liability Act (CERCLA), Resource Conservation and Recovery Act (RCRA), the Oil Pollution Act (OPA) or other state and local hazardous waste sites. The benefits from use of nanomaterials for remediation could include more rapid or cost-effective cleanup of wastes relative to current conventional approaches. Such benefits may derive from the enhanced reactivity, surface area, subsurface transport, and/or sequestration characteristics of nanomaterials.

Chloro-organics are a major class of contaminants at U.S. waste sites, and several nanomaterials have been applied to aid in their remediation. Zero-valent iron (Fig. 13) has been used successfully in the past to remediate groundwater by construction of a permeable reactive barrier (iron wall) of zero-valent iron to intercept and dechlorinate chlorinated hydrocarbons such as trichloroethylene in groundwater plumes. Laboratory studies indicate that a wider range of chlorinated hydrocarbons may be dechlorinated using various nanoscale iron particles (principally by abiotic means, with zero-valent iron serving as the bulk reducing agent), including chlorinated methanes, ethanes, benzenes, and polychlorinated biphenyls (Elliot and Zhang, 2001). Nanoscale zero-valent iron may not only treat aqueous dissolved chlorinated solvents in situ, but also may remediate the dense nonaqueous phase liquid (DNAPL) sources of these contaminants within aquifers (Quinn et al., 2005).

In addition to zero-valent iron, other nanosized materials such as metalloporphyrinogens have been tested for degradation of tetrachlorethylene, trichloroethylene, and carbon tetrachloride under anaerobic conditions (Dror, 2005). Titanium oxide based nanomaterials have also been developed for potential use in the photocatalytic degradation of various chlorinated compounds (Chen, 2005).

Enhanced retention or solubilization of a contaminant may be helpful in a remediation setting. Nanomaterials may be useful in decreasing sequestration of hydrophobic contaminants, such as polycyclic aromatic hydrocarbons (PAHs), bound to soils and sediments. The release of these contaminants from sediments and soils could make them more accessible to in situ biodegradation. For example, nanomaterials made from poly(ethylene) glycol modified urethane acrylate have been used to enhance the bioavailability of phenanthrene (Tungittiplakorn, 2005).

Metal remediation has also been proposed, using zero-valent iron and other classes of nanomaterials. Nanoparticles such as poly(amidoamine) dendrimers can serve as chelating agents, and can be further enhanced for ultrafiltration of a variety of metal ions (Cu (II), Ag(I), Fe(III), etc.) by attaching functional groups such as primary amines, carboxylates, and hydroxymates (Diallo, 2005). Other research indicates that arsenite and arsenate may be precipitated in the subsurface using zero-valent iron, making arsenic less mobile (Kanel, 2005). Self-assembled monolayers on mesoporous supports (SAMMS) are nanoporous ceramic materials that have been developed to remove mercury or radionuclides from wastewater (Mattigod, 2003).

Nanomaterials have also been studied for their ability to remove metal contaminants from air. Silica-titania nanocomposites can be used for elemental mercury removal from vapors such as those coming from combustion sources, with silica serving to enhance adsorption and titania to photocatalytically oxidize elemental mercury to the less volatile mercuric oxide (Pitoniak, 2005). Other authors have demonstrated nanostructured silica can sorb other metals generated in combustion environments, such as lead and cadmium (Lee et al., 2005; Biswas and Zachariah, 1997). Certain nanostructured sorbent processes can be used to prevent emission of nanoparticles and create byproducts that are useful nanomaterials (Biswas et al., 1998)

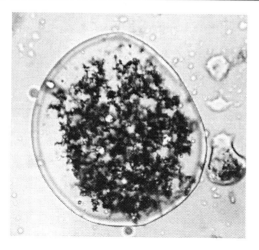

Figure 13. Nanoscale zero-valent iron encapsulated in an emulsion droplet.
These nanoparticles have been used for remdiation of sites contaminated with variuos organic pollutants. (Image cortesy of Dr. Jacqueline W. Quinn, Kennedy Space Center, NASA)

2.2.2. Sensors

Sensor development and application based on nanoscale science and technology is growing rapidly due in part to the advancements in the microelectronics industry and the increasing availability of nanoscale processing and manufacturing technologies. In general, nanosensors can be classified in two main categories: (1) sensors that are used to measure nanoscale properties (this category comprises most of the current market) and (2) sensors that are themselves nanoscale or have nanoscale components. The second category can eventually result in lower material cost as well as reduced weight and power consumption of sensors, leading to greater applicability and enhanced functionality.

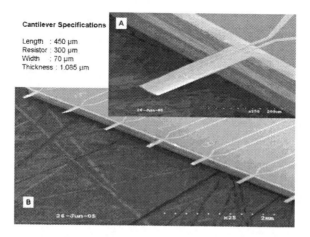

Figure 14. Piezoresistive cantilever sensor.
Devices such as these may be used to detect low levels of a wide range of substances, including pollutants, explosives, and biological or chemical warfare agents. (Image courtesy of Dr. Zhiyu Hu and Dr. Thomas Thundat, Nanoscale Science and Device Group, Oak Ridge National Laboratory)

One of the near-term research products of nanotechnology for environmental applications is the development of new and enhanced sensors to detect biological and chemical contaminants. Nanotechnology offers the potential to improve exposure assessment by facilitating collection of large numbers of measurements at a lower cost and improved specificity. It soon will be possible to develop micro- and nanoscale sensor arrays that can detect specific sets of harmful agents in the environment at very low concentrations. Provided adequate informatics support, these sensors could be used to monitor agents in real time, and the resulting data can be accessed remotely. The potential also exists to extend these small-scale monitoring systems to the individual level to detect personal exposures and *in vivo* distributions of toxicants. Figure 14 shows an example of a nanoscale sensor.

In the environmental applications field, nanosensor research and development is a relatively uncharted territory. Much of the new generation nanoscale sensor development is driven by defense and biomedical fields. These areas possess high-need applications and the resources required to support exploratory sensor research. On the other hand, the environmental measurement field is a cost sensitive arena with less available resources for leading-edge development. Therefore, environmental nanosensor technology likely will evolve by leveraging the investment in nanosensor research in related fields.

2.3. BENEFITS THROUGH OTHER APPLICATIONS THAT SUPPORT SUSTAINABILITY

Nanotechnology may be able to advance environmental protection by addressing the long-term sustainability of resources and resource systems. Listed in Table 2 are examples describing actual and potential applications relating to water, energy, and materials. Some applications bridge between several resource outcomes. For example, green manufacturing using nanotechnology (both top down and bottom up) can improve the manufacturing process by increasing materials and energy efficiency, reducing the need for solvents, and reducing waste products.

Many of the following applications can and should be supported by other agencies. However, EPA has an interest in helping to guide the work in these areas.

Table 2. Outcomes for Sustainable Use of Major Resources and Resource Systems

Water	sustain water resources of quality and availability for desired uses
Energy	generate clean energy and use it efficiently
Materials	use material carefully and shift to environmentally preferable materials
Ecosystems	protect and restore ecosystem functions, goods, and services
Land	support ecologically sensitive land management and development
Air	sustain clean and healthy air

EPA Innovation Action Council, 2005

2.3.1. Water

Nanotechnology has the potential to contribute to long-term water quality, availability, and viability of water resources, such as through advanced filtration that enables more water re-use, recycling, and desalinization. For example, nanotechnology-based flow-through capacitors (FTC) have been designed that desalt seawater using one-tenth the energy of state-of-the art reverse osmosis and one-hundredth of the energy of distillation systems. The projected capital and operation costs of FTC-based systems are expected to be one-third less than conventional osmosis systems (NNI, 2000).

Applications potentially extend even more broadly to ecological health. One long-term challenge to water quality in the Gulf of Mexico, the Chesapeake Bay, and elsewhere is the build up of nutrients and toxic substances due to runoff from agriculture, lawns, and gardens. In general with current practices, about 150% of nitrogen required for plant uptake is applied as fertilizer (Frink et al., 1996). Fertilizers and pesticides that incorporate nanotechnology may result in less agricultural and lawn/garden runoff of nitrogen, phosphorous, and toxic substances, which is potentially an important emerging application for nanotechnology that can contribute to sustainability. These potential applications are still in the early research stage (USDA, 2003). Applications involving dispersive uses of nanomaterials in water have the potential for wide exposures to aquatic life and humans. Therefore, it is important to understand the toxicity and environmental fate of these nanomaterials.

2.3.2. Energy

There is potential for nanotechnology to contribute to reductions in energy demand through lighter materials for vehicles, materials and geometries that contribute to more effective temperature control, technologies that improve manufacturing process efficiency, materials that increase the efficiency of electrical components and transmission lines, and materials that could contribute to a new generation of fuel cells and a potential hydrogen economy. However, because the manufacture of nanomaterials can be energy-intensive, it is important to consider the entire product lifecycle in developing and analyzing these technologies.

Table 3 illustrates some potential future nanotechnology contributions to energy efficiency (adapted from Brown, 2005). Brown (2005a,b) estimates that the eight technologies could result in national energy savings of about 14.5 quadrillion BTU's (British thermal units, a standard unit of energy) per year, which is about 14.5% of total U.S. energy consumption per year.

The items in Table 3 represent many different technology applications. For instance, one of many examples of molecular-level control of industrial catalysis is a nanostructured catalytic converter that is built from nanotubes and has been developed for the chemical process of styrene synthesis. This process revealed a potential of saving 50% of the energy at this process level. Estimated energy savings over the product life cycle for styrene were 8-9% (Steinfeldt et al., 2004). Nanostructured catalysts can also increase yield (and therefore reduce energy and materials use) at the process level. For example, the petroleum industry now uses nanotechnology in zeolite catalysts to crack hydrocarbons at a significantly improved process yield (NNI, 2000).

Table 3. Potential U.S. Energy Savings from Eight Nanotechnology Applications
(Adapted from Brown, 2005 a)

Nanotechnology Application	Estimated Percent Reduction in Total Annual U.S. Energy Consumption**
Strong, lightweight materials in transportation	6.2 *
Solid state lighting (such as white light LED's)	3.5
Self-optimizing motor systems (smart sensors)	2.1
Smart roofs (temperature-dependent reflectivity)	1.2
Novel energy-efficient separation membranes	0.8
Energy efficient distillation through supercomputing	0.3
Molecular-level control of industrial catalysis	0.2
Transmission line conductance	0.2
Total	**14.5**

*Assuming a 5.15 Million BTU/ Barrel conversion (corresponding to reformulated gasoline – from EIA monthly energy review, October 2005, Appendix A)

**Based on U.S. annual energy consumption from 2004 (99.74 Quadrillion Btu/year) from the Energy Information Administration Annual Energy Review 2004

There are additional emerging innovative approaches to energy management that could potentially reduce energy consumption. For example, nanomaterials arranged in superlattices could allow the generation of electricity from waste heat in consumer appliances, automobiles, and industrial processes. These thermoelectric materials could, for example, further extend the efficiencies of hybrid cars and power generation technologies (Ball, 2005).

In addition to increasing energy efficiency, nanotechnology also has the potential to contribute to alternative energy technologies that are environmentally cleaner. For example, nanotechnology is forming the basis of a new type of highly efficient photovoltaic cell that consists of quantum dots connected by carbon nanotubes (NREL, 2005). Also, gases flowing over carbon nanotubes have been shown to convert to an electrical current, a discovery with implications for novel distributed wind power (Ball, 2004).

While nanotechnology has the potential to contribute broadly to energy efficiency and cleaner sources of energy, it is important to consider energy use implications over the entire product lifecycle, particularly in manufacturing nanomaterials. Many of the manufacturing processes currently used and being developed for nanotechnology are energy intensive (Zhang et al., 2006). In addition, many of the applications discussed here are projected applications. There are still some technical and economic hurdles for these applications.

2.3.3. Materials

Nanotechnology may also lead to more efficient and effective use of materials. For example, nanotechnology may improve the functionality of catalytic converters and reduce by up to 95% the mass of platinum group metals required. This has overall product lifecycle benefits. Because platinum group metals occur in low concentration in ore, this reduction in use may reduce ecological impacts from mining (Lloyd et al., 2005). However, manufacturing precise nanomaterials can be material-intensive.

With nanomaterials' increased material functionality, it may be possible in some cases to replace toxic materials and still achieve the desired functionality (in terms of electrical conductivity, material strength, heat transfer, etc.), often with other life-cycle benefits in terms of material and energy use. One example is lead-free conductive adhesives formed from self-assembled monolayers based on nanotechnology, which could eventually substitute for leaded solder. Leaded solder is used broadly in the electronics industry; about 3900 tons lead are used per year in the United States alone. In addition to the benefits of reduced lead use, conductive adhesives could simplify electronics manufacture by eliminating several processing steps, including the need for acid flux and cleaning with detergent and water (Georgia Tech., 2005).

Nanotechnology is also used for Organic Light Emitting Diodes (OLEDs). OLEDs are a display technology substitute for Cathode Ray Tubes, which contain lead. OLEDs also do not require mercury, which is used in conventional Flat Panel Displays (Frazer, 2003). The OLED displays have additional benefits of reduced energy use and overall material use through the lifecycle (Wang and Masciangioli, 2003).

2.3.4. Fuel Additives

Nanomaterials also show potential as fuel additives and automotive catalysts and as catalysts for utility boilers and other energy-producing facilities. For example, cerium oxide nanoparticles are being employed in the United Kingdom as on- and off-road diesel fuel additives to decrease emissions (http://www.oxonica.com/cms/pressreleases/PressRelease-12-03⁻04.pdf and http://www.oxonica.com/cms/casestudies/CaseStudyV9SB.pdf). These manufacturers also claim a more than 5- 10 % decrease in fuel consumption with an associated decrease in vehicle emissions. Such a reduction in fuel consumption and decrease in emissions would result in obvious environmental benefits. Limited published research and modeling have indicated that the addition of cerium oxide to fuels may increase levels of specific organic chemicals in exhaust, and result in emission of cerium oxide (Health Effects Institute, 2001); the health impacts associated with such alterations in diesel exhaust were not examined.

In: Nanotechnology and the Environment
Editor: Robert V. Neumann

ISBN: 978-1-60692-663-5
© 2010 Nova Science Publishers, Inc.

Chapter 3

RISK ASSESSMENT OF NANOMATERIALS

U.S. Environmental Protection Agency

3.1. INTRODUCTION

Occupational and environmental exposures to a limited number of engineered nanomaterials have been reported (Baron et al., 2003; Maynard et al., 2004). Uncertainties in health and environmental effects associated with exposure to engineered nanomaterials raise questions about potential risks from such exposures (Dreher, 2004; Swiss Report Reinsurance Company, 2004; UK Royal Society Report, 2004; European Commission Report, 2004; European NanoSafe Report 2004; UK Health and Safety Executive, 2004)

EPA's mission and mandates call for an understanding of the health and environmental implications of intentionally produced nanomaterials. A challenge in evaluating risk associated with the manufacture and use of nanomaterials is the diversity and complexity of the types of materials available and being developed, as well as the seemingly limitless potential uses of these materials. A risk assessment is the evaluation of scientific information on the hazardous properties of environmental agents, the dose-response relationship, and the extent of exposure of humans or environmental receptors to those agents. The product of the risk assessment is a statement regarding the probability that humans (populations or individuals) or other environmental receptors so exposed will be harmed and to what degree (risk characterization).

EPA generally follows the risk assessment paradigm described by the National Academy of Sciences (NRC, 1983 and 1994), which at this time EPA anticipates to be appropriate for the assessment of nanomaterials (Figure 15). In addition, nanomaterials should be assessed from a life cycle perspective (Figure 16).

The overall risk assessment approach used by EPA for conventional chemicals is thought to be generally applicable to nanomaterials. It is important to note that nanomaterials have large surface areas per unit of volume, as well as novel electronic properties relative to conventional chemicals. Some of the special properties that make nanomaterials useful are also properties that may cause some nanomaterials to pose hazards to humans and the environment, under specific conditions, as discussed below. Furthermore, numerous nanomaterial coatings are being developed to enhance performance for intended applications. These coatings may impact the behavior and effects of the materials, and may or may not be retained in the environment. It will be necessary to consider these unique properties and issues, and their potential impacts on fate, exposure, and toxicity, in developing risk assessments for nanomaterials.

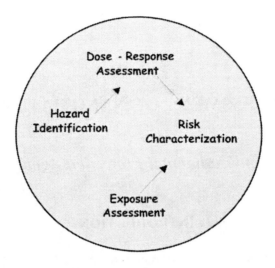

Figure 15. EPA's Risk Assessment Approach

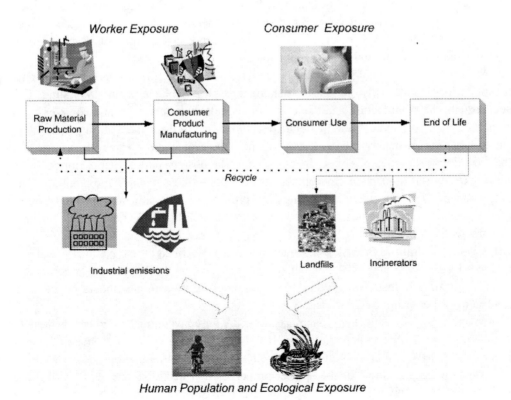

Figure 16. Life Cycle Perspective to Risk Assessment

A number of authors have reviewed characterization, fate, and toxicological information for nanomaterials and proposed research strategies for safety evaluation of nanomaterials (Morgan, 2005; Holsapple et al., 2005; Blashaw et al., 2005; Tsuji et al., 2006; Borm et al., 2006; Powers et al., 2006; Thomas and Sayre, 2005). Tsuji et al. (2006) proposed a general framework for risk assessment of nanomaterials which identified nanomaterial characteristics, such as particle size, structure/properties, coating, and particle behavior, that are expected be important in developing nanomaterial risk assessments. These issues are similar to those we note herein. Other governments have also undertaken efforts to identify research needs for nanomaterial risk assessment (UK Department for Environment, Food and Rural Affairs, 2005; Borm and Kreyling, 2004). The European Union's Scientific Committee on Emerging and Newly Identified Health Risks (SCENIHR, 2006) has also overviewed existing data on nanomaterials, data gaps, and issues to be considered in conducting risk assessments on nanomaterials.

The purpose of this chapter is to briefly review the state of knowledge regarding the components needed to conduct a risk assessment on nanomaterials. The following key aspects of risk assessment are addressed as they relate to nanomaterials: chemical identification and physical properties characterization, environmental fate, environmental detection and analysis, human exposure, human health effects, and ecological effects. Each of these aspects is discussed by providing a synopsis of key existing information on each topic.

3.2. CHEMICAL IDENTIFICATION AND CHARACTERIZATION OF NANOMATERIALS

The identification and characterization of chemical substances and materials is an important first step in assessing their risk. Understanding the physical and chemical properties in particular is necessary in the evaluation of hazard (both toxicological and ecological) and exposure (all routes). Chemical properties that are important in the characterization of discrete chemical substances include, but are not limited to, composition, structure, molecular weight, melting point, boiling point, vapor pressure, octanol-water partition coefficient, water solubility, reactivity, and stability. In addition, information on a substance's manufacture and formulation is important in understanding purity, product variability, performance, and use.

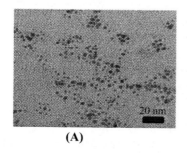 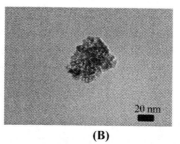

(A)   (B)

Figure 17. Transmission Electron Microscope (TEM) image of aerosol-generated TiO$_2$ nanoparticles. (A) Un-aggregated and (2-5 nm) (B) and aggregated (80-120 nm), used in exposure studies to determine the health impacts of manufactured nanoparticles. Nanoparticle aggregation may play an important role in health and environmental impacts. (Images courtesy of Vicki Grassian, University of Iowa [Grassian, et al., unpublished results])

The diversity and complexity of nanomaterials makes chemical identification and characterization not only more important but also more difficult. A broader spectrum of properties will be needed to sufficiently characterize a given nanomaterial for the purposes of evaluating hazard and assessing risk. Chemical properties such as those listed above may be important for some nanomaterials, but other properties such as particle size and size distribution, surface/volume ratio, shape, electronic properties, surface characteristics, state of dispersion/agglomeration and conductivity are also expected to be important for the majority of nanoparticles. Figure 17 provides an illustration of different states of aggregation nanoparticles. Powers et al. (2006) provides a discussion of nanoparticle properties that may be important in understanding their effects and methods to measure them.

A given nanomaterial can be produced in many cases by several different processes yielding several derivatives of the same material. For example, single-walled carbon nanotubes can be produced by several different processes that can generate products with different physical-chemical properties (e.g., size, shape, composition) and potentially different ecological and toxicological properties (Thomas and Sayre, 2005; Oberdörster et al., 2005a). It is not clear whether existing physical-chemical property test methods are adequate for sufficiently characterizing various nanomaterials in order to evaluate their hazard and exposure and assess their risk. It is clear that chemical properties such as boiling point and vapor pressure are insufficient. Alternative methods for measuring properties of nanomaterials may need to be developed both quickly and cost effectively.

Because of the current state of development of chemical identification and characterization, current chemical representation and nomenclature conventions may not be adequate for some nanomaterials. Nomenclature conventions are important to eliminate ambiguity when communicating differences between nanomaterials and bulk materials and in reporting for regulatory purposes. EPA's OPPT is participating in new and ongoing workgroup/panel deliberations with the American National Standards Institute (ANSI), the American Society for Testing and Materials (ASTM), and the International Organization for Standardization (ISO) concerning the development of terminology and chemical nomenclature for nanosized substances, and will also continue with its own nomenclature discussions with the Chemical Abstracts Service (CAS).

3.3. ENVIRONMENTAL FATE OF NANOMATERIALS

As more products containing nanomaterials are developed, there is greater potential for environmental exposure. Potential nanomaterial release sources include direct and/or indirect releases to the environment from the manufacture and processing of nanomaterials, releases from oil refining processes, chemical and material manufacturing processes, chemical clean up activities including the remediation of contaminated sites, releases of nanomaterials incorporated into materials used to fabricate products for consumer use including pharmaceutical products, and releases resulting from the use and disposal of consumer products containing nanoscale materials (e.g., disposal of screen monitors, computer boards, automobile tires, clothing and cosmetics). The fundamental properties concerning the environmental fate of nanomaterials are not well understood (European Commission, 2004), as there are few available studies on the environmental fate of nanomaterials. The following sections summarize what is known or can be inferred about the fate of nanomaterials in the atmosphere, in soils, and in water. These summaries are followed by sections discussing: 1) biodegradation, bioavailability, and bioaccumulation of nanomaterials, 2) the potential for transformation of nanomaterials to more toxic metabolites, 3) possible interactions between nanomaterials and other environmental contaminants; and 4) the applicability of current environmental fate and transport models to nanomaterials.

3.3.1. Fate of Nanomaterials in Air

Several processes and factors influence the fate of airborne particles in addition to their initial dimensional and chemical characteristics: the length of time the particles remain airborne, the nature of their interaction with other airborne particles or molecules, and the distance that they may travel prior to deposition. The processes important to understanding the potential atmospheric transport of particles are diffusion, agglomeration, wet and dry deposition, and gravitational settling. These processes are relatively well understood for ultrafine particles and may be applicable to nanomaterials as well (Wiesner et al., 2006). However, in some cases, intentionally produced nanomaterials may behave quite differently from incidental ultrafine particles, for example, nanoparticles that are surface coated to prevent agglomeration. In addition, there may be differences between freshly generated and aged nanomaterials.

With respect to the length of time particles remain airborne, particles with aerodynamic diameters in the nanoscale range (<100 nm) may follow the laws of gaseous diffusion when released to air. The rate of diffusion is inversely proportional to particle diameter, while the rate of gravitational settling is proportional to particle diameter (Aitken et al., 2004). Airborne particles can be classified by size and behavior into three general groups: Small particles (diameters <80 nm) are described as being in the agglomeration mode; they are short-lived because they rapidly agglomerate to form larger particles. Large particles (>2000 nm, beyond the discussed <100 nm nanoscale range) are described as being in the coarse mode and are subject to gravitational settling. Intermediate-sized particles (>80 nm and < 2000 nm, which includes particle sizes outside the discussed <100 nm nanoscale range) are described as being in the accumulation mode and can remain suspended in air for the longest time, days to weeks, and can be removed from air via dry or wet deposition (Bidleman, 1988; Preining, 1998; Spurny, 1998; Atkinson, 2000; UK Royal Society, 2004; Dennenkamp et al., 2002). Note that these generalizations apply to environmental conditions and do not preclude the possibility that humans and other organisms may be exposed to large as well as smaller particles by inhalation.

Deposited nanoparticles are typically not easily resuspended in the air or re-aerosolized (Colvin 2003; Aitken et al., 2004). Because physical particle size is a critical property of nanomaterials, maintaining particle size during the handling and use of nanomaterials is a priority. Current research is underway to produce carbon nanotubes that do not form clumps either by functionalizing the tubes themselves, or by treatment with a coating or dispersing agent (UK Royal Society, 2004; Colvin, 2003), so future materials may be more easily dispersed.

Many nanosized particles are reported to be photoactive (Colvin, 2003), but their susceptibility to photodegradation in the atmosphere has not been studied. Nanomaterials are also known to readily adsorb a variety of materials (Wiesner et al., 2006), and many act as catalysts. However, no studies are currently available that examine the interaction of nanosized adsorbants and chemicals sorbed to them, and how this interaction might influence their respective atmospheric chemistries.

3.3.2. Fate of Nanomaterials in Soil

The fate of nanomaterials released to soil is likely to vary depending upon the physical and chemical characteristics of the nanomaterial. Nanomaterials released to soil can be strongly sorbed to soil due to their high surface areas and therefore be immobile. On the other hand, nanomaterials are small enough to fit into smaller spaces between soil particles, and might therefore travel farther than larger particles before becoming trapped in the soil matrix. The strength of the sorption of any intentionally produced nanoparticle to soil will be dependent on its size, chemistry, applied particle surface treatment, and the conditions under which it is applied. Studies have demonstrated the differences in mobility of a variety of insoluble nanosized materials in a porous medium (Zhang, 2003; Lecoanet and Wiesner, 2004; Lecoanet et al., 2004).

Additionally, the types and properties of the soil and environment (e.g., clay versus sand) can affect nanomaterial mobility. For example, the mobility of mineral colloids in soils and sediments is strongly affected by charge (Wiesner et al., 2006). Surface photoreactions provide a pathway for nanomaterial transformation on soil surfaces. Humic substances, common constituents of natural particles, are known to photosensitize a variety of organic photoreactions on soil and other natural surfaces that are exposed to sunlight. Studies of nanomaterial transformations in field situations are further complicated by the presence of naturally occurring nanomaterials of similar molecular structures and size ranges. Iron oxides are one example.

3.3.3. Fate of Nanomaterials in Water

Fate of nanomaterials in aqueous environments is controlled by aqueous solubility or dispersability, interactions between the nanomaterial and natural and anthropogenic chemicals in the system, and biological and abiotic processes. Waterborne nanoparticles generally settle more slowly than larger particles of the same material. However, due to their high surface-area-to-mass ratios, nanosized particles have the potential to sorb to soil and sediment particles (Oberdörster et al., 2005a). Where these soil and sediment particles are subject to sedimentation, the sorbed nanoparticles can be more readily removed from the water column. Some nanoparticles will be subject to biotic and abiotic degradation resulting in removal from the water column. Abiotic degradation processes that may occur include hydrolysis and photocatalyzed reaction in surface waters. Particles in the upper layers of aquatic environments, on soil surfaces, and in water droplets in the atmosphere are exposed to sunlight. Light-induced photoreactions often are important in determining environmental fate of chemical substances.

Figure 18. Zinc oxide nanostructures synthesized by a vapor-solid process.
(Image courtesy of Prof. Zhong Win Lang of Georgia Tech.).

These reactions may alter the physical and chemical properties of nanomaterials and so alter their behavior in aquatic environments. Certain organic and metallic nanomaterials may possibly be transformed under anaerobic conditions, such as in aquatic (benthic) sediments. From past studies, it is known that several types of organic compounds are generally susceptible to reduction under such conditions. Complexation by natural organic materials such as humic colloids can facilitate reactions that transform metals in anaerobic sediments (see Nurmi et al., 2005 and references therein).

In contrast to processes that remove nanoparticles from the water column, some dispersed insoluble nanoparticles can be stabilized in aquatic environments. For example, researchers at Rice University have reported that although C60 fullerene is initially insoluble in water, it spontaneously forms aqueous colloids containing nanocrystalline aggregates. The concentration of nanomaterials in the suspensions can be as high as 100 parts per million (ppm), but is more typically in the range of 10-50 ppm. The stability of the particles and suspensions is sensitive to pH and ionic strength (CBEN, 2005; Fortner et al., 2005). Sea surface microlayers consisting of lipid, carbohydrate and proteinaceous components along with naturally-occurring colloids made up of humic acids, may have the potential to sorb nanoparticles and transport them in aquatic environments over long distances (Moore, 2006, Schwarzenbach et al., 1993). These interactions will also delay nanoparticle removal from the water column.

Heterogeneous photoreactions on metal oxide surfaces are increasingly being used as a method for drinking water, wastewater and groundwater treatment. Figure 18 shows an example of the surface of a synthesized metal oxide nanostructure, Semiconductors such as titanium dioxide and zinc oxide as nanomaterials have been shown to effectively catalyze both the reduction of halogenated chemicals and oxidation of various other pollutants, and heterogeneous photocatalysis has been used for water purification in treatment systems.

Nanoparticle photochemistry is being studied with respect to its possible application in water treatment. Processes that control transport and removal of nanoparticles in water and wastewater are being studied to understand nanoparticle fate (Moore, 2006; Wiesner et al., 2006). The fate of nanosized particles in wastewater treatment plants is not well characterized. Wastewater may be subjected to many different types of treatment, including physical, chemical and biological processes, depending on the characteristics of the wastewater, whether the plant is a publicly owned treatment work or onsite industrial facility, etc. Broadly speaking, nanosized particles are most likely to be affected by sorption processes (for example in primary clarifiers) and chemical reaction. The ability of either of these processes to immobilize or destroy the particles will depend on the chemical and physical nature of the particle and the residence times in relevant compartments of the treatment plant. As noted above, sorption, agglomeration and mobility of mineral colloids are strongly affected by pH; thus pH is another variable that may affect sorption and settling of nanomaterials. Current research in this area includes the production of microbial granules that are claimed to remove nanoparticles from simulated wastewater (Ivanov et al., 2004). Nanomaterials that escape sorption in primary treatment may be removed from wastewater after biological treatment via settling in the secondary clarifier. Normally the rate of gravitational settling of particles such as nanomaterials in water is dependent on particle diameter, and smaller particles settle more slowly. However, settling of nanomaterials could be enhanced by entrapment in the much larger sludge flocs, removal of which is the objective of secondary clarifiers.

3.3.4 Biodegradation of Nanomaterials

Biodegradation of nanoparticles may result in their breakdown as typically seen in biodegradation of organic molecules, or may result in changes in the physical structure or surface characteristics of the material. The potential for and possible mechanisms of biodegradation of nanosized particles have just begun to be investigated. As is the case for other fate processes, the potential for biodegradation will depend strongly on the chemical and physical nature of the particle. Many of the nanomaterials in current use are composed of inherently nonbiodegradable inorganic chemicals, such as ceramics, metals and metal oxides, and are not expected to biodegrade. However, a recent preliminary study found that C_{60} and C_{70} fullerenes were taken up by wood decay fungi after 12 weeks, suggesting that the fullerene carbon had been metabolized (Filley et al., 2005). For other nanomaterials biodegradability may be integral to the material's design and function. This is the case for some biodegradable polymers being investigated for use in drug transport (Madan et al., 1997; Brzoska et al., 2004), for which biodegradability is mostly a function of chemical structure and not particle size.

Biodegradability in waste treatment and the environment may be influenced by a variety of factors. Recent laboratory studies on C60 fullerenes have indicated the development of stable colloid structures in water that demonstrate toxicity to bacteria under aerobic and anaerobic conditions (CBEN, 2005; Fortner et al., 2005). Further studies are needed to determine whether fullerenes may be toxic to microorganisms under environmental conditions. One must also consider the potential of photoreactions and other abiotic processes to alter the bioavailability and thus biodegradation rates of nanomaterials. In summary, not enough is known to enable meaningful predictions on the biodegradation of nanomaterials in the environment and much further testing and research are needed.

3.3.5. Bioavailability and Bioaccumulation of Nanomaterials

Bacteria and living cells can take up nanosized particles, providing the basis for potential bioaccumulation in the food chain (Biswass and Wu, 2005). Aquatic and marine filter feeders near the base of the food chain feed on small particles, even particles down to the nanometer size fraction. The bioavailability of specific nanomaterials in the environment will depend in part on the particle. Environmental fate processes may be too slow for effective removal of persistent nanomaterials before they can be taken up by an organism. In the previous section, it was noted that some physical removal processes, such as gravitational settling, are slower for nanosized particles than for microparticles. This would lead to an increased potential for inhalation exposure to terrestrial organisms and for increased exposure of aquatic organisms to aqueous colloids. Not enough information has been generated on rates of deposition of nanomaterials from the atmosphere and surface water, or of sorption to suspended soils and sediments in the water column, to determine whether these processes could effectively sequester specific nanoparticles before they are taken up by organisms.

Complexation of metallic nanomaterials may have important interactive effects on biological availability and photochemical reactivity. For example, the biological availability of iron depends on its free ion concentrations in water and the free ion concentrations are affected by complexation. Complexation reduces biological availability by reducing free metal ion concentrations and dissolved iron is quantitatively complexed by organic ligands. Solar UV radiation can interact with these processes through photoreactions of the complexes. Further, iron and iron oxides can participate in enzymatic redox reactions that change the oxidation state, physical chemical properties and bioavailability of the metal (Reguera et al., 2005).

3.3.6. Potential for Toxic Transformation Products from Nanomaterials

Certain nanomaterials are being designed for release as reactants in the environment, and therefore are expected to undergo chemical transformation. One example of this is iron (Fe^0) nanoparticles employed as reactants for the dechlorination of organic pollutants (Zhang, 2003). As the reaction progresses, the iron is oxidized to iron oxide. Other metal particles are also converted to oxides in the presence of air and water. Whether the oxides are more or less toxic than the free metals depends on the metal. Under the right conditions, certain metal compounds could be converted to more mobile compounds. In these cases, small particle size would most likely enhance this inherent reactivity. Some types of quantum dots have been shown to degrade under photolytic and oxidative conditions, and furthermore, compromise of quantum dot coatings can reveal the metalloid core, which may be toxic (Hardman, 2006). Degradation products from carbon nanomaterials (fullerenes and nanotubes) have not yet been reported.

3.3.7. Interactions between Nanomaterials and Organic or Inorganic Contaminants: Effects and the Potential for Practical Applications

The examples cited in this section illustrate how nanomaterials have been demonstrated to alter the partitioning behavior of chemicals between environmental compartments and between the environment and living organisms. Furthermore, several nanomaterials are reactive toward chemicals in the environment, generate reactive species, or catalyze reactions of other chemicals. These properties are currently under study for use in waste remediation operations. It should be noted that the potential also exists for nanomaterials to effect unforseen changes, if released to the environment in large quantities.

Two types of effects under study for possible exploitation are sorption and reaction. The high surface area of nanosized particles provides enhanced ability to sorb both organic and inorganic chemicals from environmental matrices compared to conventional forms of the same materials. This property can potentially be utilized to bind pollutants to enhance environmental remediation. Many examples of immobilized nanomaterials for use in pollution control or environmental remediation have been described in the literature. These include nanosponges or nanoporous ceramics, large particulate or bead materials with nanosized pores or crevasses (Christen, 2004), and solid support materials with coatings of nanoparticles (for example, see Comparelli et al., 2004). This section will instead focus on releases of free nanoparticles and effects on chemicals in the environment. The remainder of this section will be organized into known changes in the mobility of chemicals caused by their sorption to nanoparticles, and known instances of reactivity and catalytic activity toward chemicals mediated by nanoparticles.

No single overall effect can be described for the sorption of chemicals to nanomaterials based on their size or chemical makeup alone. In air, aerosolized nanoparticles can adsorb gaseous or particulate pollutants. In soil or sediments, nanomaterials might increase the bioavailability of pollutants, thereby increasing the pollutant=s availability for biodegradation (UK Royal Society, 2004). Depending on the conditions, nanosized carbon such as C_{60} or nanotubes could either enhance or inhibit the mobility of organic pollutants (Cheng et al., 2004). Stable colloids of hydrophobic nanomaterials in an aqueous environment could provide a hydrophobic microenvironment that suspends hydrophobic contaminants and retards their rate of deposition onto soils and sediments. Similar effects are known to happen with naturally occurring colloids made up of humic acids and suspended sediment particles (Schwarzenbach et al., 1993). Nanoparticles can be altered to optimize their affinities for particular pollutants by modifying the chemical identity of the polymer.

Several studies investigating the sorption of organic pollutants and metals in air, soil, and water to nanosized materials have recently been reported in the literature. The sorption of naphthalene to C_{60} from aqueous solution was compared to sorption to activated carbon (Cheng et al., 2004). The investigators observed a correlation between the surface area of the particles and the amount of naphthalene adsorbed from solution. In other studies, nanoparticles made of an amphiphilic polymer have been shown to mobilize phenanthrolene from contaminated sandy soil and increase its bioavailability (Tungittiplakorn et al., 2004, 2005). It has been reported that magnetite crystals adsorb arsenic and chromium (CrVI) from water (CBEN, 2005; Hu et al., 2004), suggesting potential purification techniques for metal-laden drinking water (CBEN, 2005). The adsorption and desorption of volatile organic compounds from ambient air by fullerenes has been investigated (Chen et al., 2000). Inhalation exposures of benzo(a)pyrene sorbed to ultrafine aerosols of Ga_2O_3 (Sun et al., 1982) and diesel exhaust (140 nm) (Sun et al., 1984) were studied in rats. The studies showed that when compared to inhalation of pure benzo(a)pyrene aerosols, material sorbed to the gallium oxide had increased retention in the respiratory tract, and increased exposure to the stomach, liver, and kidney.

Nanoscale materials are typically more reactive than larger particles of the same material. This is true especially for metals and certain metal oxides. In the environment, nanomaterials have the potential to react with a variety of chemicals; their increased or novel reactivity coupled with their sorptive properties allows for accelerated removal of chemicals from the environment. Many groups are currently investigating the use of nanomaterials for the destruction of persistent pollutants in the environment.

Nanoscale iron particles have been demonstrated to be effective in the in situ remediation of soil contaminated with tetrachloroethylene. A wide variety of additional pollutants are claimed to be transformed by iron nanoparticles in laboratory experiments, including halogenated (Cl, Br) methanes, chlorinated benzenes, certain pesticides, chlorinated ethanes, polychlorinated hydrocarbons, TNT, dyes, and inorganic anions such as nitrate, perchlorate, dichromate, and arsenate. Further investigations are underway with bimetallic nanoparticles (iron nanoparticles with Pt, Pd, Ag, Ni, Co, or Cu deposits) and metals deposited on nanoscale support materials such as nanoscale carbon platelets and nanoscale polyacrylic acid (Zhang, 2003). Nanosized clusters of C_{60} have been shown to generate reactive oxygen species in water under UV and polychromatic light. Similar colloids have been reported to degrade organic contaminants and act as bacteriocides (Boyd et al., 2005). Fullerol ($C_{60}(OH)_{24}$) has also been demonstrated to produce reactive oxygen species under similar conditions (Pickering and Wiesner, 2005).

3.3.8. Applicability of Current Environmental Fate and Transport Models to Nanomaterials

When performing exposure assessments on materials for which there are no experimental data, models are often used to generate estimated data, which can provide a basis for making regulatory decisions. It would be advantageous if such models could be applied to provide estimated properties for nanomaterials, since there is very little experimental data available for these materials. The models used by EPA's Office of Pollution Prevention and Toxics (OPPT) to assess environmental fate and exposure, are, for the most part, designed to provide estimates for organic molecules with defined and discrete structures. These models are not designed for use on inorganic materials; therefore, they cannot be applied to inorganic nanomaterials. Many models derive their estimates from structural information and require that a precise structure of the material of interest be provided. Since many of the nanomaterials in current use, such as quantum dots, ceramics and metals, are solids without discrete molecular structures, it is not possible to provide the precise chemical structures that these models need. While it is usually possible to determine distinct structures for fullerenes, the models cannot accept the complex fused-ring structures of the fullerenes. Also, the training sets of chemicals with which the quantitative structure-activity relationships (QSAR) in the models were developed do not include fullerene-type materials. Fullerenes are unique materials with unusual properties, and they cannot be reliably modeled by QSARs developed for other substantially different types of materials.

In general, models used to assess the environmental fate and exposure to chemicals are not applicable to intentionally produced nanomaterials. Depending on the relevance of the chemical property or transformation process, new models may have to be developed to provide estimations for these materials; however, models cannot be developed without the experimental data needed to design and validate them. Before the environmental fate, transport and multimedia partitioning of nanomaterials can be effectively modeled, reliable experimental data must be acquired for a variety of intentionally produced nanomaterials.

However, models are also used which focus on the fate and distribution of particulate matter (air models) and/or colloidal materials (soil, water, landfill leachates, ground water), rather than discrete organics. For example, fate of atmospheric particulate matter (e.g., PM_{10}) has been the subject of substantial research interest and is a principal regulatory focus of EPA's Office of Air and Radiation. Since intentionally produced nanomaterials are expected to be released to and exist in the environment as particles in most cases, it is wise to investigate applicability of these other models. In fact it can be reasoned that the most useful modeling tools for exposure assessment of nanomaterials are likely to be found not in the area of environmental fate of specific organic compounds (more precisely, prediction of their transport and transformation), rather in fields in which the focus is on media-oriented pollution issues: air pollution, water quality, ground water contamination, etc. A survey of such tools should be made and their potential utility for nanomaterials assessed.

3.4. ENVIRONMENTAL DETECTION AND ANALYSIS OF NANOMATERIALS

The challenge in detecting nanomaterials in the environment is compounded not only by the extremely small size of the particles, but also by their unique physical structure and physico-chemical characteristics. The varying of physical and chemical properties can significantly impact the extraction and analytical techniques that can be used for the analysis of a specific nanomaterial. As noted above, the chemical properties of particles at the nanometer size can significantly differ from the chemical properties of larger particles consisting of the same chemical composition. Independent of the challenges brought on by the intrinsic chemical and physical characteristics of nanomaterials, the interactions of nanomaterials with and in the environment, including agglomeration, also provide significant analytical challenges. Some nanomaterials are being developed with chemical surface treatments that maintain nanoparticle properties in various environments. These surface treatments can also complicate the detection and analysis of nanomaterials.

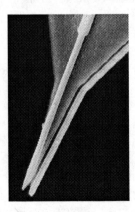

Figure 19. SEM of a scanning gate probe.
The large tip is the probe for a scanning tunneling microscope, and the smaller is a gate that allows sharper imaging of the sample. Instruments such as these can be used to analyze nanomaterials. (Image courtesy of Prof. Leo Kouwehnhoven, Delft University of Technology. Reprinted with permission from Gurevich, L., et al., 2000) (Copyright 2000, American Institute of Physics.)

In characterizing an environmental sample for intentionally produced nanomaterials, one must be able to distinguish between the nanoparticles of interest and other ultra-fine particles, such as nanoscale particles in the atmosphere generated from coal combustion or forest fires, or nanoscale particles in aquatic environments derived from soil runoff, sewage treatment, or sediment resuspension. Information used to help characterize nanomaterials includes particle size, morphology, surface area and chemical composition. Other information taken into consideration in identifying the source of nanomaterials includes observed particle concentrations mapped over an area along with transport conditions (e.g., meteorology, currents) at the time of sampling. For nanomaterials with unique chemical composition as found in some quantum dots containing heavy metals, chemical characterization (qualitative and quantitative) can play an important role in their detection and source identification.

The level of effort needed and costs to perform analysis for nanomaterials will depend on which environmental compartment samples are being taken from, as well as the type of desired analytical information. The analysis of nanomaterials from an air matrix requires significantly less (if any) "sample" preparation than samples taken from a soil matrix where it is necessary to employ greater efforts for sample extraction and/or particle isolation. Analytical costs also depend on the degree of information being acquired. Analyzing samples for number concentration (i.e., the number and size distribution of nanoparticles per unit volume) requires significantly less effort than broadening such analyses to include characterization of particle types (fullerenes, quantum dots, nanowires, etc.). The level of effort also increases for elemental composition analyses.

Although significant advances in aerosol particle measurement technology have been made over the past two decades in response to National Ambient Air Quality Standards (U.S. EPA, 2004), many of these technologies were designed to effectively function on micron sized particles, particles hundreds to a thousand times larger than nanoparticles, and are not effective in the separation or analysis of particles at the nanometer scale. However, some of these technologies have advanced so that they are effective in providing separation and analytical data relevant to nanoparticles.

The information available from the bulk analysis of nanomaterials from environmental samples has limitations when one is trying to identify a specific nanomaterial. As stated previously, nanoscale particles generated by natural and other anthropogenic sources cannot be separated from nanomaterials of interest using sampling methodologies based upon particle size. During analysis, detected signals generated by nanoscale particles that are not of interest can mask or augment the signals of nanomaterials of interest, resulting in inadequate or erroneous data. Where procedures are available for the selective extraction of nanomaterials of interest, one can avoid interfering signals from other nanoscale particles obtained during sampling. In the case of inseparable mixtures of natural and engineered/manufactured nanomaterials, the use of single particle analysis methodologies may be necessary to provide definitive analysis for the engineered/manufactured nanomaterials.

Even given all the challenges presented in analyzing for specific nanomaterials of interest, methods and technologies are available that have demonstrated success. For aerosols, multi-stage impactor samplers are available commercially that can separate and collect nanoparticle size fractions for subsequent analysis. These technologies provide nanoparticle fraction separation based upon the aerodynamic mobility properties of the particles. Aerodynamic mobility-based instruments include micro-orifice uniform deposit impactors (MOUDIs), and electrical low-pressure impactors (ELPIs) (McMurry, 2000). There are also aerosol fractionation and collection technologies based upon the electrodynamic mobility of particles. These technologies use the mobility properties of charged nanoparticles in an electrical field to obtain particle size fractionation and collection. Instruments employing this technology include differential mobility analyzers (DMAs) and scanning mobility particle sizers (SMPSs) (McMurry, 2000).

Available technologies for the size fractionation and collection of nanoparticle fractions in liquid mediums include size-exclusion chromatography, ultrafiltration and field flow fractionation (Powers et al., 2006; Rocha et al., 2000; Willis, 2002; Chen and Selegue, 2002). On-line particle size analysis in liquid mediums can be done using various techniques including dynamic light scattering (DLS) to obtain a particle size distribution (Biswas and Wu, 2005) and inductively-coupled mass spectrometry (ICP-MS), a technique that provides chemical characterization information (Chen and Beckett, 2001). For more definitive analytical data, single-particle analytical techniques can be employed. Single-particle laser microprobe mass spectrometry (LAMMS) can provide chemical composition data on single particles from a collected fraction (McMurry, 2000). Electron microscopy techniques [e.g., transmission electron microscopy (TEM), scanning electron microscopy (SEM)] can provide particle size, morphological and chemical composition information on collected single nanoparticles in a vacuum environment. Figure 19 shows an SEM of a scanning gate probe, which is an example of an instrument that can be used to analyze nanomaterials. Atomic Force Microscopy, a relatively new technology, can provide particle size and morphological information on single nanoparticles in liquid, gas, and vacuum environments (Maynard, 2000).

3.5. HUMAN EXPOSURES AND THEIR MEASUREMENT AND CONTROL

As the use of nanomaterials in society increases, it is reasonable to assume that their presence in environmental media will increase proportionately, with consequences for human and environmental exposure. Potential human exposures to nanomaterials, or mixtures of nanomaterials, include workers exposed during the production, use, recycling and disposal of nanomaterials, general population exposure from releases to the environment as a result of the production, use, recycling and disposal in the workplace, and direct general population exposure during the use of commercially available products containing nanomaterials. This section identifies potential sources, pathways, and routes of exposure, discusses potential means for mitigating or minimizing worker exposure, describes potential tools and models that may be used to estimate exposures, and identifies potential data sources for these models.

3.5.1. Exposure to Nanomaterials

The exposure paradigm accounts for a series of events beginning from when external mechanisms (e.g., releases or handling of chemicals) make a chemical available for absorption or other mode of entry at the outer body boundary to when the chemical or its metabolite is delivered to the target organ. Between outer body contact with the chemical and delivery to the target organ, a chemical is absorbed and distributed. Depending on the nature of the chemical and the route of exposure, the chemical may be metabolized. For the purposes of this section, we will limit the discussion to the types of resources that are needed (and available) to assess exposure up to the point where it is distributed to the target organ.

3.5.2. Populations and Sources of Exposure

The potential for intentionally produced nanomaterials to be released into the environment or used in quantities that raise human exposure concerns are numerous given their predicted widespread applications in products. This section discusses some of the potential sources and pathways by which humans may be exposed to nanomaterials.

3.5.2.1. Occupational Exposure

Workers may be exposed to nanoscale materials during manufacturing/synthesis of the nanoscale materials, during formulation or end use of products containing the nanoscale material, or during disposal or recycling of the products containing the nanoscale materials. Because higher concentrations and amounts of nanoscale materials and higher frequencies and exposures are more likely in workplace settings, occupational exposures warrant particular attention.

Table 4 presents the potential sources of occupational exposure during the common methods for nanoscale material synthesis: gas phase synthesis, vapor deposition, colloidal, and attrition methods.

While there are several potential exposure sources for each manufacturing process, packaging, transfer, and cleaning operations may provide the greatest potential for airborne levels of nanomaterials and resultant occupational exposures. "The risk of particle release during production seems to be low, because most production processes take place in closed systems with appropriate filtering systems. Contamination and exposure to workers is more likely to happen during handling and bagging of the material and also during cleaning operations." (Luther, 2004).

During the formulation of the nanomaterials into products (e.g., coatings and composite materials), workplace releases and exposures may be most likely to occur during the transfer/unloading of nanoscale material from shipping containers and during cleaning of process equipment and vessels. During the use of some of these products in workplace settings, releases of and exposures to nanoscale material are highly dependent upon the application. For example, workers who manually apply spray coatings often have higher levels of occupational exposure. Alternately, components of composites are usually bound in the composite matrix, and workers handling the composites would generally have lower levels of occupational exposure. Exposure could also occur during product machining (e.g., cutting, drilling and grinding), repair, destruction and recycling [National Institute for Occupational Health and Safety (NIOSH), 2005a]. NIOSH (2004, 2005b) has issued additional documents on nanotechnology and workplace safety and associated research needs.

3.5.2.2. Release and General Population Exposure

General population exposure may occur from environmental releases from the production and use of nanomaterials and from direct use of products containing nanomaterials. During the production of nanomaterials, there are several potential sources for environmental releases including the evacuation of production chambers, filter residues, losses during spray drying, emissions from filter or scrubber break-through, and wastes from equipment cleaning and product handling. No data have been identified quantifying the releases of nanomaterials from industrial processes or of the fate of nanomaterials after release into the environment. However, due to the small size of nanomaterials, they will likely stay airborne for a substantially longer time than other types of particulate. The most likely pathway for general population exposure from releases from industrial processes is direct inhalation of materials released into the air during manufacturing (UK Royal Society, 2004). Releases from industrial or transportation accidents, natural disasters, or malevolent activity such as a terrorist attack may also lead to exposure of workers or the general public.

Table 4. Potential Sources of Occupational Exposure for Various Synthesis Methods (adapted from Aitken, 2004)

Synthesis Process	Particle Formation	Exposure Source or Worker Activity	Primary Exposure Route
Gas Phase	in air	Direct leakage from reactor, especially if the reactor is operated at positive pressure.	Inhalation
		Product recovery from bag filters in reactors.	Inhalation/ Dermal
		Processing and packaging of dry powder.	Inhalation/ Dermal
		Equipment cleaning/maintenance (including reactor evacuation and spent filters).	Dermal (and Inhalation during reactor evacuation)
Vapor Deposition	on substrate	Product recovery from reactor/dry contamination of workplace.	Inhalation
		Processing and packaging of dry powder.	Inhalation / Dermal
		Equipment cleaning/maintenance (including reactor evacuation).	Dermal (and Inhalation during reactor evacuation)
Colloidal/ Attrition	liquid suspension	If liquid suspension is processed into a powder, potential exposure during spray drying to create a powder, and the processing and packaging of the dry powder.	Inhalation / Dermal
		Equipment cleaning/maintenance.	Dermal

Note: Ingestion would be a secondary route of exposure from all sources/activities from deposition of nanomaterials on food or mucous that is subsequently swallowed (primary exposure route inhalation) and from hand-to-mouth contact (primary exposure route dermal). Ocular exposure would be an additional route of exposure from some sources/activities from deposition of airborne powders or mists in the eyes or from splashing of liquids.

Table 5. Examples of Potential Sources of General Population and/or Consumer Exposure for Several Product Types

Product Type	Release and/ or Exposure Source	Exposed Population	Potential Exposure Route
Sunscreen containing nanoscale material	Product application by consumer to skin	Consumer	Dermal
	Release by consumer (e.g., washing with soap and water) to water supply	General population	Ingestion
	Disposal of sunscreen container (with residual sunscreen) after use (to landfill or incineration)	General population	Inhalation or Ingestion
Metal catalysts in gasoline for reduce-ing vehicle exhaust*	Release from vehicle exhaust to air (then deposition to surface water)	General population	Inhalation or Ingestion
Paints and Coatings	Weathering, disposal	Consumers, general population,	Dermal, Inhalation or Ingestion
Clothing	Wear, washing, disposal	Consumers, general population	Dermal, inhalation, ingestion from surface or groundwater
Electronics	Release at end of life or recycling stage	Consumers, general population	Dermal, ingestion from surface or groundwater
Sporting goods	Release at end of life or recycling stage	Consumers, general population	Dermal, inhalation, ingestion from surface or groundwater

Note: This is not an exhaustive list of consumer products or exposure scenarios. Ingestion would be a secondary route of exposure from some sources from deposition of nanomaterials on food or mucous that is subsequently swallowed (primary exposure route inhalation) and from hand-to-mouth contact (primary exposure route dermal). Ocular exposure would be an additional route of exposure from some sources/activities from deposition of airborne powders or mists in the eyes or from splashing of liquids.

* Metal catalysts are not currently being used in gasoline in the U.S. Cerium oxide nanoparticles are being marketed in Europe as on and off-road diesel fuel additives.

Nanoscale materials have potential applications in many consumer products resulting in potential general population exposure. Electronics, medicine, cosmetics, chemistry, and catalysis are potential beneficiaries of nanotechnology. Widespread exposure via direct contact with products from these sectors is expected. Table 5 presents several examples of potential sources of general population and consumer exposure associates with the use of such products.

3.5.3. Exposure Routes

Much remains to be scientifically demonstrated about the mechanisms by which human exposure to nanomaterials can occur. Intentionally produced nanomaterials share a number of characteristics, such as size and dimensions, with other substances (e.g., ultrafine particles) for which a large body of information exists on how they access the human body to cause toxicity. The data from these other substances focus primarily on inhalation as the route of exposure. However, as the range of applications of nanomaterials expands, other routes of exposure, such as dermal, ocular, and oral, may also be found to be significant in humans.

3.5.3.1. Inhalation Exposure

A UK Health and Safety Executive reference suggests that aerosol science would be applicable to airborne nanoparticle behavior. Aerosol behavior is primarily affected by particle size and the forces of inertia, gravity, and diffusion. Other factors affecting nanoparticle airborne concentrations are agglomeration, deposition, and re-suspension. (UK Health and Safety Executive, 2004) All of these issues, which are discussed in more detail in the reference, are relevant for understanding, predicting, and controlling airborne concentrations of nanomaterials.

One reference study was found to have investigated issues involved with aerosol release of a single-walled carbon nanotube (SWCNT) material. This study noted that while laboratory studies indicate that sufficient agitation can release fine particles into the air, aerosol concentrations of SWCNT generated while handling unrefined material in the field at the work loads and rates observed were very low on a mass basis (Maynard et al., 2004). The study suggests that more research will be needed in this area.

3.5.3.2. Ingestion Exposure

Information on exposure to nanoscale environmental particles via oral exposure is lacking. In addition to traditional ingestion of food, food additives, medicines and dietary supplements, dust and soil (particularly in the case of children), ingestion of inhaled particles can also occur (such as through the activities of the mucocilliary escalator). The quantity ingested is anticipated to be relatively small in terms of mass.

3.5.3.3. Dermal Exposure

Dermal exposure to nanomaterials has received much attention, perhaps due to concerns with occupational exposure and the introduction of nanomaterials such as nanosized titanium dioxide into cosmetic and drug products. One reference study was found to have investigated issues involved with potential dermal exposure to a SWCNT material. The study suggests that more research will be needed in this area. This study noted that airborne particles of SWCNT may contribute to potential dermal exposure along with surface deposits due to material handling. Surface deposits on gloves were estimated to be between 0.2 mg and 6 mg per hand. (Maynard et al., 2004)

There is an ongoing debate over the potential for penetration through "healthy/intact" versus damaged skin. Hart (2004) highlights physiological characteristics of the skin that may permit the absorption of nanosized materials. In particular the review highlights a conceivable route for the absorption of nanoparticles as being through interstices formed by stacking and layering of the calloused cells of the top layer of skin (Hart, 2004). Movement through these interstices will subsequently lead to the skin beneath, from which substances can be absorbed into the blood stream. Nanomaterials also have a greater risk of being absorbed through the skin than macro-sized particles (Tinkle, 2003). Reports of toxicity to human epidermal keratinocytes in culture following exposure to carbon nanotubes have been made (Shvedova et al., 2003; Monteiro-Riviere et al., 2005). A significant amount of intradermally injected nanoscale quantum dots were found to disperse into the surrounding viable subcutis and to draining lymph nodes via subcutaneous lymphatics (Roberts, D.W. et al., 2005). It has recently been reported that quantum dots with different physicochemical properties (size, shapes, coatings) penetrated the stratum corneum and localized within the epidermal and dermal layers of intact porcine skin within a maximum 24 hours of exposure (Ryman-Rasmussen et al., 2006). Drug delivery studies using model wax nanoparticles have provided evidence that nanoparticle surface charge alters blood-brain barrier integrity and permeability (Lockman et al., 2004).

3.5.3.4. Ocular Exposure
Ocular exposure to nanomaterials has received little attention. However, the potential for ocular exposure to nanomaterials from deposition of airborne powders or mists in the eyes or from splashing of liquids must also be considered.

3.5.4. Exposure Mitigation

Approaches exist to mitigate exposure to fine and ultrafine particulates. Some approaches such as engineering controls are applicable to the work place and may mitigate environmental releases while others such as personal protective equipment (PPE) are primarily applicable to the workplace. NIOSH suggests considering the range of control technologies and personal protective equipment demonstrated to be effective with other fine and ultrafine particles (NIOSH, 2005a). In the hierarchy of exposure reduction methods, engineering controls are preferred over PPE.

3.5.4.1. Engineering Controls
Engineering controls, and particularly those used for aerosol control, should generally be effective for controlling exposures to airborne nanoscale materials (NIOSH, 2005a). Depending on particle size, nanoparticles may diffuse rapidly and readily find leakage paths in engineering control systems in which containment is not complete (Aitken et al., 2004). However, a well-designed exhaust ventilation system with a high efficiency particulate air (HEPA) filter should effectively remove nanoparticles (Hinds, 1999). As with all filters, the filter must be properly seated to prevent nanoparticles from bypassing the filter, decreasing the filter efficiency (NIOSH, 2003). Aitken et al. (2004) recommends that engineering controls (e.g., enclosures, local exhaust ventilation, fume hoods) used to control exposure to nanoparticles need to be of similar quality and specification as those typically used for gases. However, the report also notes that no research has been identified evaluating the effectiveness of engineering controls for nanoparticles.

Efficient ultrafine particle control devices (e.g., soft x-ray enhanced electrostatic precipitation systems) may have applicability to nanoparticles control (Kulkarni et al., 2002). HEPA filters may be effective, and validation of their effectiveness is currently being studied (NIOSH, 2005a). Magnetic filter systems in welding processes have proven effective in capturing magnetic oxides and the use of nanostructured sorbents in smelter exhausts to prepare ferroelectric materials may also have applicability (Biswas et al., 1998).

3.5.4.2. Personal Protective Equipment (PPE)

Properly fitted respirators with a HEPA filter may be effective at removing nanomaterials. Contrary to intuition, fibrous filters trap smaller and larger particles more effectively than mid-sized particles. Small particles (<100 nm) tend to make random Brownian motions due to their interaction with gas molecules. The increased motion causes the particle to "zig-zag around" and have a greater chance of hitting and sticking to the fiber filter (Luther, 2004). Intermediate-sized particles (>80 nm and < 2000 nm) can remain suspended in air for the longest time. (Bidleman, 1988; Preining, 1998; Spurny, 1998; Atkinson, 2000; UK Royal Society, 2004; Dennenkamp et al., 2002)

NIOSH certifies particulate respirators by challenging them with sodium chloride (NaCl) aerosols with a count median diameter 75 nm or dioctyl phthalate (DOP) aerosols with a count median diameter of 185 nm [42 CFR Part 84.181(g)], which have been found to be in the most penetrating particle size range (Stevens and Moyer, 1989). However, as with all respirators, the greatest factor in determining their effectiveness is not penetration through the filter, but rather the face-seal leakage bypassing the device. Due to size and mobility of nanomaterials in the air, leakage may be more prevalent although no more than expected for a gas (Aitken, 2004). Only limited data on face-seal leakage has been identified. Work done by researchers at the U.S. Army RDECOM on a headform showed that mask leakage (i.e., simulated respirator fit factor) measured using submicron aerosol challenges (0.72 μm polystyrene latex spheres) was representative of vapor challenges such as sulfur hexafluoride (SF6) and isoamyl acetate (IAA) (Gardner et al., 2004).

PPE may not be as effective at mitigating dermal exposure. PPE is likely to be less effective against dermal exposure to nanomaterials than macro-sized particles from both human causes (e.g., touching face with contaminated fingers) and PPE penetration (Aitken, 2004). However, no studies were identified that discuss the efficiency of PPE at preventing direct penetration of nanomaterials through PPE or from failure due to human causes.

3.5.5. Quantifying Exposure to Nanomaterials

There is broad consensus that mass dose alone is insufficient to characterize exposure to nanomaterials (Oberdörster et al., 2005a, b; NIOSH, 2005a, b). Many studies have indicated that toxicity increases with decreased particle size and that particle surface area is a better metric for measuring exposures (Aitken, 2004). This is of particular concern for nanomaterials, which typically have very high surface-area-to-mass ratios. Additionally, there currently are no convenient methods for monitoring the surface area of particles in a worker's breathing zone or ambient air. While there could be a correlation between mass and surface area, large variations in particle weight and surface area can occur within a given batch. The average particle weight and average particle surface area of the nanomaterials being assessed would also be required for any assessments based on surface area. (Maynard and Kuempel, 2005). It has also been recommended that mass, surface area, and particle number all be measured for nanomaterials (Oberdörster et al., 2005b).

3.5.6. Tools for Exposure Assessment

Several tools exist for performing exposure assessments including monitoring data, exposure models, and the use of analogous data from existing chemicals. The following sections discuss these tools and their potential usefulness in assessing exposure to nanoscale materials.

3.5.6.1. Monitoring Data

Types of monitoring data that can be used in exposure assessment include biological monitoring, personal sampling, ambient air monitoring, worker health monitoring and medical surveillance. Although monitoring and measurement are discussed earlier in section3.4, the discussion below includes coverage of some issues directly pertinent to exposure.

Biological Monitoring Biomonitoring data, when permitted and applied correctly, provides the best information on the dose and levels of a chemical in the human body. Examples of bio-monitoring include the Centers for Disease Control and Prevention (CDC) national monitoring program and smaller surveys such as the EPA's National Human Exposure Assessment Survey (NHEXAS). Biomonitoring can be the best tool for understanding the degree and spread of exposure, information that cannot be captured through monitoring concentrations in ambient media. Biomonitoring, however, is potentially limited in its application to nanotechnology because it is a science that is much dependent on knowledge of biomarkers, and its benefits are highest when there is background knowledge on what nanomaterials should be monitored. Given the current limited knowledge on nanoscale materials in commerce, their uses, and their fate in the environment and in the human body, it is difficult to identify or prioritize nanomaterials for biomonitoring. Should biomonitoring become more feasible in the future, it presents an opportunity to assess the spatial and temporal distribution of nanomaterials in workers and the general population.

Personal Sampling Personal sampling data provide an estimate of the exposure experienced by an individual, and can be an important indicator of exposure in occupational settings. It is limited in that it does not account for changes to the dose received by the target organ after the biological processes of absorption, distribution, metabolism and excretion. Generally, for cost and feasibility reasons, personal and biomonitoring data are not available for all chemicals on a scale that is meaningful to policymakers. Also, the applicability of personal sampling to nanomaterials is dependent on the development of tools for accurately detecting and measuring such materials in ambient media.

Ambient Monitoring Ambient media monitoring measures concentrations in larger spaces such as in workplaces, homes or the general environment. Ambient data are used as assumed exposure concentrations of chemicals in populations when it is not feasible or practical to conduct personal sampling for individuals in the populations. Typically, these data are used in models in addition to other assumptions regarding exposure parameters, including population activities and demographics such as age.

Challenges of Monitoring As discussed in Section 3.4, there are many challenges to detecting and characterizing nanoscale materials, including the extremely small size of the analyte, as well as the need to distinguish the material of interest from other similarly-sized materials, the tendency for nanoparticles to agglomerate, and the cost of analysis. Additionally, as discussed in above, it is not always clear what the most appropriate metric is to measure. Mass may not be the most appropriate dose metric; therefore, techniques may be required for measuring particle counts and surface area, or other parameters. These problems are compounded when there is a need for monitoring data to be used in exposure assessment. Monitoring equipment should be not only sensitive and specific, but also easy to use, durable, able to operate in a range of environments, and affordable. Additionally, data sometimes needs to be collected continuously and analyzed in real-time. Further, the nanomaterials may need to be measured in a variety of media and several properties may need to be measured in parallel. All of the current measuring methods and instruments individually fall short of adequately addressing all of these needs.

3.5.6.2. Exposure Modeling
A recent use of ambient monitoring data to estimate the exposure of a population is the cumulative exposure project for air toxics recently completed for hazardous air toxics using the Hazardous Air Pollutant Exposure Model (HAPEM) (http://www.epa.gov/ttn/fera/ human_hapem.html). This model predicts inhalation exposure concentrations of air toxics from all outdoor sources, based on ambient concentrations from modeling or monitor data for specific air toxics at the census tract level.

Other EPA screening level models include the Chemical Screening Tool for Exposures and Environmental Releases (ChemSTEER) (http://www.epa.gov/oppt/exposure/docs/chems teer.htm) and the Exposure and Fate Assessment Screening Tool (E-FAST) (http://www.epa.gov/oppt/exposure/docs/efast.htm). ChemSTEER estimates potential dose rates for workers and environmental releases from workplaces. E-FAST uses the workplace releases to estimate potential dose rates for the general population. E-FAST also estimates potential dose rates for consumers in the general public. However, whether ChemSTEER and E-FAST will be useful for assessments of nanoscale materials is not clear because of the significantly different chemical and physical properties of nanomaterials.

Challenges of Using Models with Nanoscale Materials There are several models that span multiple levels of complexity and are designed to estimate exposure at several points in the exposure paradigm. The effectiveness of these models at predicting human exposure will depend on the parameters and assumptions of each model. For models that are based on assumptions specific to the chemical such as the physical and chemical properties, and interactions in humans and the environment based on these properties, much substance-specific data may be required.

Data Sets for Modeling The availability of ambient data is clearly critical to modeling exposure, and there are a number of resources within EPA for this type of data. In some cases such as for pesticides, the exposure can be anticipated based on the quantity of the substance that is proposed to be applied and the anticipated residue on a food item as an example. Sometimes there are data collected under statutory obligations, such as data collected for the Toxics Release Inventory (TRI) under the Emergency Planning and Community Right to Know Act (EPCRA). For contaminants in drinking water, the data may be reported to the Safe Drinking Water Information System (SDWIS). Generating data for nanomaterials necessitates the identification of nanomaterials as separate and different from other chemicals of identical nomenclature, and their classification as toxic substances, or in a manner that adds nanomaterials to the list of reportable releases/contaminants.

Though not fully representative of population exposure, workplace data have frequently provided the foundation for understanding exposure and toxicity for many chemicals in industrial production. A recent study in the United States, in which ambient air concentrations and glove deposit levels were measured, identified a concern for exposure during handling of nanotubes (Maynard et al., 2004). In the work environment, data on workplace exposure is frequently collected under the purview of Occupational Safety and Health Administration (OSHA)-mandated programs to assess worker exposure and assure compliance with workplace regulations and worker protection. Employers, however, are not required to report these data. In addition, OSHA standards are typically airborne exposure levels that are based on health or economic criteria or both, and typically only defined exceedences of these standards are documented. To understand nanotechnology risks in the workplace, the National Institute of Occupational Safety and Health (NIOSH) is advancing initiatives to investigate amongst other issues, nanoparticle exposure and ways of controlling exposure in the workplace (NIOSH, 2004).

3.6. HUMAN HEALTH EFFECTS OF NANOMATERIALS

There is a significant gap in our knowledge of the environmental, health, and ecological implications associated with nanotechnology (Dreher, 2004; Swiss Report, 2004; UK Royal Society, 2004; European NanoSafe, 2004; UK Health and Safety Executive, 2004). This section provides an overview of currently available information on the toxicity of nanoparticles; much of the information is for natural or incidentally formed nanosized materials, and is presented to aid in the understanding of intentionally produced nanomaterials.

3.6.1. Adequacy of Current Toxicological Database

The Agency's databases on the health effects of particulate matter (PM), asbestos, silica, or other toxicological databases of similar or larger sized particles of identical chemical composition (U.S. EPA, 1986, 1996, 2004) should be evaluated for their potential use in conducting toxicological assessments of intentionally produced nanomaterials. The toxicology chapter of the recent *Air Quality Criteria for Particulate Matter document* cites hundreds of references describing the health effects of ambient air particulate matter including ultrafine ambient air ($PM_{0.1}$), silica, carbon, and titanium dioxide particles (U.S.

EPA, 2004). However, it is important to note that ambient air ultrafine particles are distinct from intentionally produced nanomaterials since they are not purposely engineered and represent a physicochemical and dynamic complex mixture of particles derived from a variety of natural and combustion sources. In addition, only approximately five percent of the references cited in the current *Air Quality Criteria for Particulate Matter* document describe the toxicity of chemically defined ultrafine particles, recently reviewed by Oberdörster et al. (2005a) and Donaldson et al. (2006).

A search of the literature on particle toxicity studies published up to 2005 confirms the paucity of data describing the toxicity of chemically defined ultrafine particles and, to an even greater extent, that of intentionally produced nanomaterials (Figure 20). The ability to assess the toxicity of intentionally produced carbon nanotubes by extrapolating from the current carbon-particle toxicological database was examined by Lam et al. (2004) and Warheit et al. (2004). Their findings demonstrate that graphite is not an appropriate safety reference standard for carbon nanotubes, since carbon nanotubes displayed very different mass-based dose-response relationships and lung histopathology when directly compared with graphite.

These initial findings indicate a high degree of uncertainty in the ability of current particle toxicological databases to assess or predict the toxicity of intentionally produced carbon-based nanomaterials displaying novel physicochemical properties. Additional comparative toxicological studies are needed to assess the utility of the current particle toxicological databases in assessing the toxicity of other classes or types of intentionally produced nanomaterials, as well as to relate their health effects to natural or anthropogenic ultrafine particles.

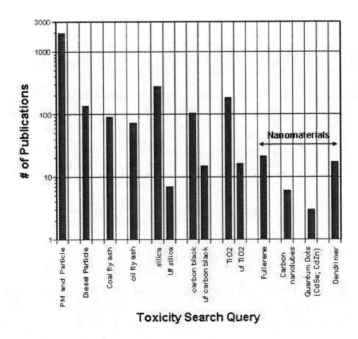

Figure 20. Particle Toxicology Citations.
Results depict the number of toxicological publications for each type of particle obtained from a PubMed search of the literature up to 2005 using the indicated descriptors and the term "toxicity." Uf denotes ultrafine size (<100nm) particles.

3.6.2. Toxicity and Hazard Identification of Engineered/Manufactured Nanomaterials

Studies assessing the role of particle size on toxicity have generally found that ultrafine or nanosize range (<100 nm) particles are more toxic on a mass-based exposure metric when compared to larger particles of identical chemical composition (Oberdörster et al., 1994; Li et al., 1999; Höhr et al., 2002). Other studies have shown that particle surface area dose is a better predictor of the toxic and pathologic responses to inhaled particles than is particle mass dose (Oberdörster et al., 1992; Driscoll, 1996; Lison et al., 1997; Donaldson et al., 1998; Tran et al., 2000; Brown et al., 2001; Duffin et al., 2002). Studies examining the pulmonary toxicity of carbon nanotubes have provided evidence that intentionally produced nanomaterials can display unique toxicity that cannot be explained by differences in particle size alone (Lam et al., 2004; Warheit et al., 2004). For example, Lam reported single walled carbon nanotubes displayed greater pulmonary toxicity than carbon black nanoparticles. Similar results have been obtained from comparative *in vitro* cytotoxicity studies (Jia et al., 2005). Muller et al. (2005) reported multi-walled carbon nanotubes to be more proinflammatory and profibrogenic when compared to ultrafine carbon black particles on an equivalent mass dose metric. Shvedova et al. (2005) reported unusual inflammatory and fibrogenic pulmonary responses to specific nanomaterials, suggesting that they may injure the lung by new mechanisms. Exposure of human epidermal keratinocyte cells in culture to single-walled carbon nanotubes was reported to cause dermal toxicity, including oxidative stress and loss of cell viability (Shvedova et al., 2003). The combination of small particle size, large surface area, and ability to generate reactive oxygen species have been suggested as key factors in induction of lung injury following exposure to some incidentally produced nanomaterials (Nel et al., 2006).

Contrary to other reports, Uchino et al. (2002), Warheit et al. (2006) and Sayes et al. (2006) have reported nanoscale titanium dioxide toxicity was not found to be dependent on particle size and surface area. These authors reported that specific crystal structure and the ability to generate reactive oxygen species are important factors to consider in evaluating nanomaterial toxicity. Similar to other reports, Warheit demonstrated that nanomaterial coating impacted toxicity (Warheit et al., 2005).

Studies have demonstrated that nanoparticle toxicity is extremely complex and multi-factorial, potentially being regulated by a variety of physicochemical properties such as size and shape, as well as surface properties such as charge, area, and reactivity (Sayes et al., 2004; Cai et al., 1992; Sclafani and Herrmann, 1996; Nemmar et al., 2003; Derfus et al., 2004). The properties of carbon nanotubes in relation to pulmonary toxicology have recently been reviewed (Donaldson et al., 2006).

Toxicological assessment of intentionally produced nanomaterials will require information on the route (inhalation, oral, dermal) that carries the greatest risk for exposure to these materials, as well as comprehensive physicochemical characterization of them in order to provide information on size, shape, as well as surface properties such as charge, area, and reactivity. Establishment of dose-response relationships linking physicochemical properties of intentionally produced nanomaterials to their toxicities will identify the appropriate exposure metrics that best correlate with adverse health effects.

One of the most striking findings regarding particle health effects is the ability of particles to generate local toxic effects at the site of initial deposition as well as very significant systemic toxic responses (U.S. EPA, 2004). Pulmonary deposition of polystyrene nanoparticles was found to not only elicit pulmonary inflammation but also to induce vascular thrombosis (Nemmar et al., 2003). Pulmonary deposition of carbon black nanoparticles was found to decrease heart rate variability in rats and prolonged cardiac repolarization in young healthy individuals in recent toxicological and clinical studies (Harder et al., 2005; Frampton et al., 2004). Extrapulmonary translocation following pulmonary deposition of carbon black nanoparticles was reported by Oberdörster et al. (2004a, 2005a) Submicron particles have been shown to penetrate the stratum corneum of human skin following dermal application, suggesting a potential route by which the immune system may be affected by dermal exposure to nanoparticles (Tinkle et al., 2003; Ryman-Rasmussen et al., 2006). Zhao et al. (2005) have reported that in molecular dynamic computer simulations C_{60} fullerenes bind to double and single-stranded DNA and note that these simulations suggest that C_{60} may negatively impact the structure, stability, and biological functions of DNA. It is clear that toxicological assessment of intentionally produced nanomaterials will require consideration of both local and systemic toxic responses (e.g., immune, cardiovascular, neurological toxicities) in order to ensure that that we identify the health effects of concern from these materials.

3.6.3. Adequacy of Toxicity Test Methods for Nanomaterials

A challenge facing the toxicological assessment of intentionally produced nanomaterials is the wide diversity and complexity of the types of materials that are available commercially or are under development. In many cases, the same type of nanomaterial can be produced by several different processes, giving rise to a number of versions of the same type of nanomaterial. For example, single-walled carbon nanotubes can be mass produced by four different processes, each of which generates products of different size, shape, composition, and potentially different toxicological properties (Bekyarova, 2005). It is not known whether the toxicological assessment of one type and source of nanomaterial will be sufficient to assess the toxicity of the same class/type of nanomaterial produced by a different process. Manufactured materials may also be treated with coatings, or other surface modifications, in order to generate mono-dispersed suspensions that extend and enhance their unique properties. The extent to which surface modifications of intentionally produced nanomaterials affect their toxicity is not known. Other testing issues include the possibility of physicochemical changes in the material before and after administration in a test system, presenting a challenge in identifying the characterization criteria for nanomaterial toxicity. Test methods that determine the toxicity and hazardous physicochemical properties of intentionally produced nanomaterials in an accepted, timely and cost effective manner are needed in order provide health risk assessment information for the diversity of such nanomaterials that are currently available (Oberdörster et al., 2005b).

3.6.4. Dosimetry and Fate of Intentionally Produced Nanomaterials

Much of what is known regarding particle dosimetry and fate has been derived from pulmonary exposure studies using ultrafine metal oxide and carbon black studies (U.S. EPA,

2004; Oberdörster, 1996; Oberdörster et al., 2005a, b; Oberdörster et al., 2004a; Kreyling et al., 2002). Ultrafine carbon black and metal oxide particles display differential deposition patterns within the lung when compared to larger sized particles of identical chemical composition. For example, 1 nm particles are preferentially deposited in the nasopharyngeal region while 5nm particles are deposited throughout the lung and 20 nm particles are preferentially deposited in the distal lung within the alveolar gas exchange region (Oberdörster et al., 2005a). Host susceptibility factors such as pre-existing lung disease significantly affect the amount and location of particles deposited within the lung. For example, individuals with chronic obstructive pulmonary disease have 4-fold higher levels of particles deposited in their upper bronchioles when compared to health individuals exposed to the same concentration of particles (U.S. EPA, 2004). Also, pulmonary deposited ultrafine particles can evade the normal pulmonary clearance mechanisms and translocate by a variety of pathways to distal organs (Oberdörster et al. 2004a, 2005a; Kreyling et al., 2002; Renwick et al., 2001). Additional studies that provide information on the deposition and fate of inhaled nanomaterials include studies in animals (Takenaka et al., 2001; Oberdörster et al., 2002) and studies in humans (Brown et al., 2002; Chalupa et al., 2004).

The deposition and fate of the class of nanomaterials called dendrimers have been examined to some degree due to their potential drug delivery applications (Malik et al 2000; Nigavekar et al., 2004.). Both studies demonstrated the critical role which surface charge and chemistry play in regulating the deposition and clearance of dendrimers in rodents.

A significant amount of intradermally injected nanoscale quantum dots were found to disperse into the surrounding viable subcutis and to draining lymph nodes via subcutaneous lymphatics (Roberts, D.W. et al., 2005). Other studies (Tinkle et al., 2003) have shown enhanced penetration of submicron fluorospheres into the stratum corneum of human skin following dermal application and mechanical stimulation. Drug delivery studies using model wax nanoparticles have provided evidence that nanoparticle surface charge alters blood-brain barrier integrity and permeability (Lockman et al., 2004). It has recently been reported that quantum dots with different physicochemical properties (size, shapes, coatings) penetrated the stratum corneum and localized within the epidermal and dermal layers of intact porcine skin within a maximum 24 hours of exposure (Ryman-Rasmussen et al., 2006). A recent review noted that quantum dots cannot be considered a uniform group of substances, and that size, charge, concentration, coating, and oxidative, photolytic, and mechanical stability are determining factors in quantum dot toxicity as well as their absorption, distribution, metabolism and excretion (Hardman, 2006). Toxicological studies have demonstrated the direct cellular uptake of multi-walled carbon nanotubes by human epidermal keratinocytes (Monteiro-Riviere et al., 2005).

Very little is known regarding the deposition and fate of other types or classes of intentionally produced nanomaterials following inhalation, ingestion, or dermal exposures. Knowledge of tissue and cell specific deposition, fate and persistence of engineered or manufactured nanomaterials, as well as factors such as host susceptibility and nanoparticle physicochemical properties regulating their deposition and fate, is needed to determine exposure-dose-response relationships associated with various routes of exposures. Information on the fate of nanomaterials is needed to assess their persistence in biological systems, a property that regulates accumulation of these particles to levels that may produce adverse health effects following long-term exposures to low concentrations of these particles.

At a 2004 nanotoxicology workshop at the University of Florida (Roberts, S.M., 2005), concerns were expressed about the ability of existing technologies to detect and quantify intentionally produced nanomaterials in biological systems. New detection methods or approaches, such as the use of labeled or tagged nanomaterials, may have to be developed in order to analyze and quantify nanomaterials within biological systems.

3.6.5. Susceptible Subpopulations

Particle toxicology research has shown that not all individuals in the population respond to particle exposures in the same way or to the same degree (U.S. EPA, 2004). Host susceptibility factors that influence the toxicity, deposition, fate and persistence of intentionally produced nanomaterials are largely unknown, although a study regarding the deposition of nanoparticles in the respiratory tract of asthmatics has been published (Chalupa et al., 2004). More information is critically needed to understand the exposure-dose-response relationships of intentionally produced nanomaterials in order to recommend safe exposure levels that protect the most susceptible subpopulations.

3.6.6. Health Effects of Environmental Technologies That Use Nanomaterials

The potential for adverse health effects may arise from direct exposure to intentionally-produced nanomaterials and/or byproducts associated with their applications. Nanotechnology is being employed to develop pollution control and remediation applications. Reactive zero-valent iron nanoparticles are being used to treat soil and aquifers contaminated with halogenated hydrocarbons, such as TCE (trichloroethylene) or DCE (dichloroethylene), and heavy metals (www.bioxtech.com). However, the production of biphenyl and benzene associated with nanoscale zero-valent iron degradation of more complex polychlorinated hydrocarbons has been reported (Elliott et al., 2005).

Photocatalytic titanium dioxide nanoparticles (nano-TiO_2) are being incorporated into building materials such as cement and surface coatings in order to reduce ambient air nitrogen oxides (NOx) levels. The European Union Photocatalytic Innovative Coverings Applications for Depollution Assessment has evaluated the effectiveness of photocatalytic nano-TiO_2 to decrease ambient air NOx levels and has concluded that this technology represents a viable approach to attain 21 ppb ambient air NOx levels in Europe by 2010 (www.picada-project.com). However, the extent to which nano-TiO_2 reacts with other ambient air co-pollutants and alters their corresponding health effects is not known.

Cerium oxide nanoparticles are being employed in the United Kingdom as on- and off-road diesel fuel additives to decrease emissions and some manufacturers are claiming fuel economy benefits. However, one study employing a cerium additive with a particulate trap has shown cerium to significantly alter the physicochemistry of diesel exhaust emissions resulting in increased levels of air toxic chemicals such as benzene, 1,3-butadiene, and acetaldehyde. Modeling estimates have predicted that use of a cerium additive in diesel fuel would significantly increase the ambient air levels of cerium (Health Effects Institute, 2001). The health impacts associated with these alterations in diesel exhaust have not been examined and are currently not known.

Environmental technologies using nanotechnology lead to direct interactions of reactive, intentionally produced nanomaterials with chemically complex mixtures present within a variety of environmental media such as soil, water, ambient air, and combustion emissions. The health effects associated with these interactions are unknown. Research will be needed to assess the health and environmental risks associated with environmental applications of nanotechnology.

3.7. ECOLOGICAL EFFECTS OF NANOMATERIALS

Nanomaterials may affect aquatic or terrestrial organisms differently than larger particles of the same materials. As noted above, assessing nanomaterial toxicity is extremely complex and multi-factorial, and is potentially influenced by a variety of physicochemical properties such as size and shape, and surface properties such as charge, area, and reactivity. Furthermore, use of nanomaterials in the environment may result in novel byproducts or degradates that also may pose risks. The following section summarizes available information and considerations regarding the potential ecological effects of nanomaterials.

3.7.1. Uptake and Accumulation of Nanomaterials

Based on analogy to physical-chemical properties of larger molecules of the same material, it may be possible to estimate the tendency of nanomaterials to cross cell membranes and bioaccumulate. However, current studies have been limited to a very small number of nanomaterials and target organisms. Similarly, existing knowledge could lead us to predict a mitigating effect of natural materials in the environment (e.g., organic carbon); however, this last concept would need to be tested for a wide range of intentionally produced nanomaterials.

Molecular weight (MW) and effective cross-sectional diameter are important factors in uptake of materials across the gill membranes of aquatic organisms or the GI tract of both aquatic and terrestrial organisms. Uptake via passive diffusion of neutral particles is low, but still measurable within a range of small molecular weights (600-900) (Zitko, 1981; Opperhuizen et al., 1985; Niimi and Oliver, 1988; McKim et al., 1985). The molecular weight of some nanomaterials falls within this range. For example, the MW of n-C60 fullerene is about 720, although the MW of a C84 carbon nanotube is greater than 1000. Passive diffusion through gill membranes or the GI tract also depends on the cross sectional diameter of particles (Opperhuizen et al., 1985; Zitko, 1981). Existing evidence indicates that the absolute limit for passive diffusion through gills is in the nanometer range (between 0. 95 and 1.5 nm), which suggests that passive diffusion may be possible for nanomaterials within this range, but not for nanomaterials with larger effective cross-sectional diameters.

Charge is also an important characteristic to consider for nanomaterial uptake and distribution. For example, as noted above, drug delivery studies using model wax nanoparticles have provided evidence that nanoparticle surface charge alters blood-brain barrier integrity and permeability in mammals (Lockman et al., 2004).

Other chemical and biotic characteristics may need to be considered when predicting accumulation and toxicity of nanoparticles in aquatic systems. For example, the Office of Water uses several specific characteristics, including water chemistry (e.g., dissolved organic carbon and particulate organic carbon) and biotic (lipid content and trophic level) characteristics, when calculating national bioaccumulation factors for highly hydrophobic neutral organic compounds (U.S. EPA, 2003).

Because the properties of some nanomaterials are likely to result in uptake and distribution phenomena different from many conventional chemicals, it is critically important to conduct studies that will provide a solid understanding of these phenomena with a range of nanomaterials and species. Studies related to human health effects assessment will provide an important foundation for understanding mammalian exposures and some cross-species processes (e.g., ability to penetrate endothelium and move out of the gut and into the organism). However, other physiology differs among animal classes, most notably respiratory physiology (e.g., gills in aquatic organisms and air sacs and unidirectional air flow in birds), while plants and invertebrates (terrestrial and aquatic) have even greater physiological differences. Because of their size, the uptake and distribution of nanomaterials may follow pathways not normally considered in the context of conventional materials (e.g., pinocytosis, facilitated uptake, and phagocytosis).

3.7.2. Aquatic Ecosystem Effects

To date, very few ecotoxicity studies with nanomaterials have been conducted. Studies have been conducted on a limited number of nanoscale materials, and in a limited number of aquatic species. There have been no chronic or full life-cycle studies reported.

For example, Oberdörster (2004b) studied effects of fullerenes in the brain of juvenile largemouth bass and concluded that C60 fullerenes induce oxidative stress, based on their observations that (a) there was a trend for reduced lipid peroxidation in the liver and gill, (b) significant lipid peroxidation was found in brains, and (c) the metabolic enzyme glutathione-S¬transferease (GST) was marginally depleted in the gill. However, no concentration-response relationship was evident as effects observed at a low dose were not observed at the single higher dose and no changes in fish behavior were observed; effects could have been due to random variation in individual fish.

Oberdörster (2004c) tested uncoated, water soluble, colloidal fullerenes (nC_{60}) and estimated a Daphnid 48-hour LC_{50} (forty-eight-hour concentration that was lethal for 50 percent of the animals in the test) at 800 parts per billion (ppb), using standard EPA protocols. Lovern and Klaper (2006) tested titanium dioxide (TiO_2) and uncoated C_{60} fullerenes in an EPA standard, 48-hour acute toxicity test using Daphnia magna. Toxicity of titanium dioxide particles and fullerenes differed by an order of magnitude, with fullerene particle solutions (particle clumps measured as 10-20 nm diameter) having an LC_{50} of 460 ppb and titanium dioxide (10-20 nm) with an LC50 of 5.5 parts per million (ppm). Particle preparation impacted toxicity: filtering solutions to remove particles larger than 100 nm resulted in LC50 of 7.9 ppm, while larger titanium dioxide clumps yielded no measurable toxicity. Large particles Figure 21 shows nanoparticles in the gut and lipid storage droplets of Daphnia magna following uptake from water.

Additionally, in behavior tests with filtered fullerenes, *Daphnia* exhibited behavioral responses, with juveniles showing an apparent inability to swim down from the surface and adults demonstrating sporadic swimming and disorientation (Lovern and Klaper, 2005). Further research on ecological species is clearly needed.

Toxicity studies and structure-activity relationship predictions for carbon black and suspended clay particles, based on analyses by EPA's OPPT, suggest that some suspended natural nanosized particles in the aquatic environment will have low toxicity to aquatic organisms, with effects thresholds ranging from tens to thousands of parts per million. Limited preliminary work with engineered/manufactured nanomaterials seems to substantiate this conclusion. For example, Cheng and Cheng (2005) reported that aggregates of single-walled carbon nanotubes (SWCNT) added to zebrafish embryos reduced hatching rate at 72 hrs, but by 77 hrs post fertilization all embryos in the treated group had hatched. However, when evaluating a limited data set of nanoscale materials (i.e., carbon black and clay only), available information on differences in toxicity observed between natural and engineered or manufactured nanomaterials should be considered. For example, as noted previously, SWCNTs displayed greater pulmonary toxicity than carbon black nanoparticles (Lam et al., 2004). Shvedova et al. (2005) reported unusual inflammatory responses to specific nanomaterials in mammals, suggesting that some nanomaterials may injure organs by novel mechanisms.

Recent reports suggest that nanomaterials may be effective bactericidal agents against both gram positive and negative bacteria in growth media (Fortner et al., 2005). The ability of these "nano-C_{60}" aggregates to inhibit the growth and respiration of microbes needs to be demonstrated under more realistic conditions. For example, effects on microbes in sewage sludge effluent and natural communities of bacteria in natural waters should be examined.

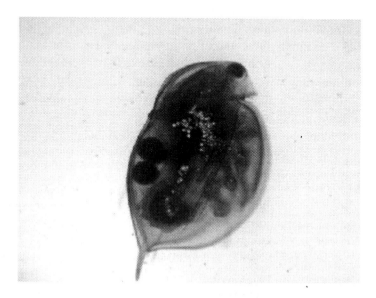

Figure 21. Fluorescent nanoparticles in water flea (*Daphnia magna*).
Adult and neonate *Daphnia* were exposed to 20nm and 1000nm fluorescently tagged carboxylated nanospheres for up to 24 hours. Nanoparticles were observed in gut and fatty lipid storage droplets using laser scanning confocal microscopy. (Image courtesy of Teresa Fernandes and Philipp Rosenkranz, Copyright Napier University. Research funded by CSL [DEFRA, UK])

3.7.3. Terrestrial Ecosystem Effects

To date, very few studies have successfully been conducted to assess potential toxicity of nanomaterials to ecological terrestrial test species (plants, wildlife, soil invertebrates, or soil microorganisms).

For terrestrial mammals, toxicity test data on rats and mice obtained for human health risk assessments should be considered. For example, studies described above indicate that ultrafine or nanosize range particles are more toxic on a mass-based exposure metric when compared to larger particles of identical chemical composition in studies of lung toxicity (Oberdörster et al., 1994; Li et al., 1999; Höhr et al., 2002), and some nanomaterials can display unique toxicity that cannot be explained by differences in particle size alone (Lam et al., 2004; Warheit et al., 2004). Toxicity to mammalian epidermal cell in culture has also been reported (Shvedova et al., 2003).

The same properties of nanomaterials that regulate uptake in aquatic organisms may limit uptake of nanoparticles by plant roots or transport through plant leaves and stomata (i.e., reducing passive transport at lower MW or size). Additionally, because many nanomaterials are designed to have strongly reactive surfaces, it is quite possible that significant pathways for toxicity may exist without uptake (e.g., disruption of respiratory epithelium structure/function or other surface cell structure/function). In a recent study of nanomaterial effects on plants, Yang and Watts (2005), reported that alumina nanoparticles (13 nm) slowed root growth in a soil-free exposure medium. Species tested included commercially important species used in ecological risk assessments of pesticides: corn (Zea mays), cucumber (Cucumis sativus), soybean (Glycine max), cabbage (Brassica oleracea), and carrot (Daucus corota). The authors reported that coating the alumina nanoparticles with an organic compound (phenanthrene), reduced the nanomaterial's effect of root elongation inhibition. Larger alumina particles (200-300 nm) did not slow root growth, indicating that the alumina itself was not causing the toxicity. The authors hypothesized that the surface charge on the alumina nanoparticles may have played a role in the decreased plant root growth. It should be noted that these studies were conducted in Petri dishes without soil, so environmental relevance is uncertain. Further, Murashov (2006) noted some limitations of this chapter including lack of discussion of known phytotoxicity of alumina, and that the increased solubility of nanoscale alumina may have resulted in increased concentrations of alumina species, which may have contributed to the observed phytotoxicity, as opposed to the nanoscale properties of the alumina.

Fundamentally, our ability to extrapolate toxicity information from conventional substances to nanomaterials will require knowledge about uptake, distribution, and excretion rates as well as modes of toxic action, and may be informed by existing structure-activity relationships (SARs), such as SARs for polycationic polymers, published in Boethling and Nabholz (1997). Synthesis of radio-labeled nanomaterials (e.g., carbon-14 labeled nanotubes) may be a useful tool, along with advanced microscopy (e.g., comparable to techniques used for asbestos quantification) for developing information on sites of toxic action and metabolic distribution.

3.7.4. Ecological Testing Issues

Because nanomaterials are often engineered to have very specific properties, it seems reasonable to presume that they may end up having unusual toxicological effects. Experiences with conventional chemicals suggest that in these cases, chronic effects of exposure are often a more important component of understanding ecological risk than acute lethality. As such, initial studies should include longer-term exposures measuring multiple, sub-lethal endpoints. They should be conducted (using appropriate forms and routes of exposure) in a manner that will elucidate key taxonomic groups (i.e., highly sensitive organisms that may become indicator species) and endpoints that may be of greatest importance to determining ecological risk. These studies must also include careful tracking of uptake and disposition to understand toxicity as a function of dose at the site of action.

A number of existing test procedures that assess long-term survival, growth, development, and reproductive endpoints (both whole organism and physiological or biochemical) for invertebrates, fish, amphibians, birds, and plants (including algae, rooted macrophytes, and terrestrial plants) should be adaptable to nanomaterials. These tests are able to examine a wide range of species and endpoints to help pinpoint the types of effects most significant to the evaluation of nanomaterials, and have a strong foundation relative to projecting likely ecological effects. Both pilot toxicity testing protocols and definitive protocols should be evaluated with respect to their applicability to nanomaterials. In addition, field studies or mesocosm studies might be conducted in systems known to be exposed to nanomaterials to screen for food chain bioaccumulation and unanticipated effects or endpoints.

In: Nanotechnology and the Environment   ISBN: 978-1-60692-663-5
Editor: Robert V. Neumann   © 2010 Nova Science Publishers, Inc.

Chapter 4

RESPONSIBLE DEVELOPMENT

U.S. Environmental Protection Agency

One of the stated goals of the National Nanotechnology Initiative is to support responsible development of nanotechnology. EPA administers a statutory framework laid out in this chapter that supports responsible development. EPA also funds and conducts research, and identifies research needs within the context of its programmatic statutory mandates. The ways that risks are characterized and decisions are made vary based on the program area (air, water, chemical substances, etc.) and also the specific statute involved (for example, Clean Air Act, Clean Water Act, Toxic Substances Control Act). Supporting responsible development at EPA is informed by an understanding of the risk from exposure to potential hazard. Section 4 of this paper discusses the risk assessment process and the types of information that EPA could need to inform its decisions. Figure 22 identifies EPA office roles, statutory authorities, and categories of research needs related to nanotechnology. As illustrated in Figure 22 and described in greater detail in Chapter 5, an understanding of environmental applications, chemical identification, potential environmental release, environmental fate and transport, human exposure and mitigation, human and environmental effects, risk assessments, and pollution prevention is needed to provide sound scientific information that informs the responsible development of nanotechnology.

4.1. RESPONSIBLE DEVELOPMENT OF NANOSCALE MATERIALS

EPA recognizes the potential benefits of nanomaterials. To fully realize that potential, the responsible development of such products is in the interest of EPA, state environmental protection agencies, producers, their suppliers, as well as users of nanotechnology, and society as a whole. EPA believes that a proactive approach is appropriate in responsible development. EPA believes that partnerships with industrial sectors will ensure that responsible development is part of initial decision making. Working in partnership with producers, their suppliers, and users of nanomaterials to 'develop best practices and standards in the workplace, throughout the supply chain, as well as other environmental programs, would help ensure the responsible development of the production, use, and end of life management of nanomaterials.

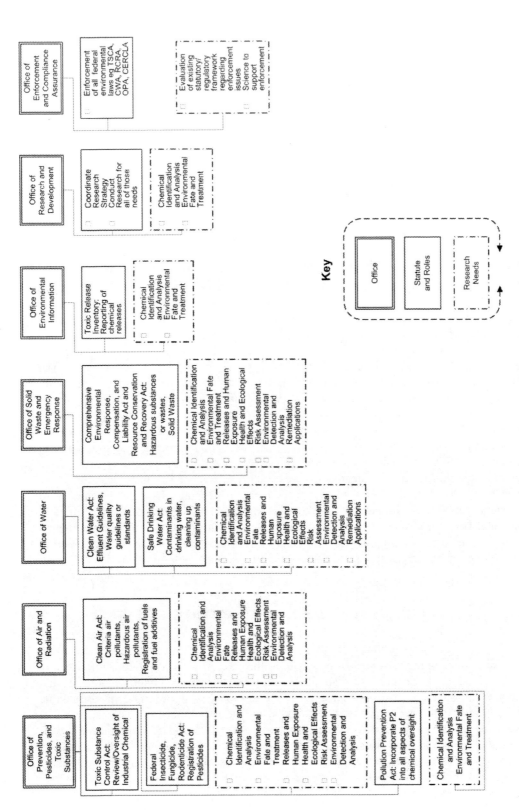

Figure 22. EPA Office Roles, Statutory Authorities, and Categories of Research Needs Related to Nanotechnology.

Responsible development of nanomaterials may present issues that are not easily characterized because of the breadth of categories of such substances. Some nanoscale materials are produced under established industrial hygiene practices based on their history of manufacturing processes and use. Human and environmental exposure information for these particular substances likely would already be available to inform responsible development. For some other nanoscale materials, there is less understanding of expected exposure and potential hazard. The uncertainty may be greater where new industrial methods are employed.

EPA intends to review as appropriate new nanotechnology products and processes as they are introduced, under EPA's product review authorities, such as TSCA, FIFRA, and the Clean Air Act. EPA intends to work with producers and users of nanomaterials to develop protocols and approaches that ensure responsible development. As new knowledge becomes incrementally available through the research needs identified in this white paper, refinement of approaches may be needed.

4.2. PROGRAM AREAS

EPA administers a wide range of environmental statutes, some of which may apply to nanomaterials depending on the specific media of application or release, such as air or water. Other statutes may apply to certain nanomaterials depending on their specific uses, applications, and processes and may require EPA to evaluate the nanomaterials before they enter into commerce (such as pesticides, fuel additives, etc.). Some risk management activities carried out under these statutes could also utilize nanomaterials as products for environmental remediation or pollution prevention technologies. The statutes administered by EPA outlined below are a starting point for evaluating and managing risks and benefits from nanomaterials. Some current EPA policies and regulations may require modifications to address this new technology.

Nanoscale materials will present other novel risk assessment/management challenges. Standards that need to be developed include terminology/nomenclature, material characterization, metrology, testing procedures, and detection methodology. There is also a need to review conventional hazard, exposure, and risk assessment tools for their applicability to nanomaterials, as well as development of risk mitigation options that are tailored to nanoscale materials. There may also be a need to review and modify reporting tools under various statutes to best cover nanoscale materials.

4.2.1. Chemical Substances

Generally, nanoscale materials that meet the definition of "chemical substances" under the Toxic Substances Control Act (TSCA), but which are not on the TSCA Inventory, must be reported to EPA according to section 5(a) of the Act, which provides for pre-manufacture review. The premanufacture review process serves as a gatekeeper to identify concerns and exercise appropriate regulatory oversight. For example, use restrictions, occupational exposure limits/controls, limits on releases to the environment and limits on manufacture may be required until toxicity and fate data are developed to better inform a risk assessment of the chemical. As previously noted EPA already is reviewing premanufacture notifications for some nanomaterials that have been received under TSCA. EPA also may review under section 5(a) of TSCA nanomaterials that represent significant new uses of chemicals already on the TSCA Inventory.

Under TSCA, EPA has the authority, by rule, to prohibit or limit the manufacture, import, processing, distribution in commerce, use, or disposal of a chemical substance; require development of test data; and/or require reporting of health and safety studies, categories of use, production volume, byproducts, an estimate of the number of individuals potentially exposed, and duration of such exposures, if the necessary findings or determinations are made. Nanomaterials that meet the definition of a chemical substance under TSCA could be subject to some or all of these provisions and programs.

4.2.2. Pesticides

Under the Federal Insecticide, Fungicide and Rodenticide Act (FIFRA), EPA is responsible for registering pesticide products for distribution or sale in the United States. An application for registration under FIFRA must disclose to EPA the specific chemicals in the pesticide formulation. Pesticide registration decisions are based on a detailed assessment of the potential effects of a product on human health and the environment, when used according to label directions. FIFRA requires EPA and states to establish programs to protect workers, and to provide training and certification for applicators. Pesticide products containing nanomaterials will be subject to FIFRA's review and registration requirements. In addition, to the extent that the use of pesticide products containing nanomaterials results in residues in food, the resulting residues require the establishment of a tolerance (maximum allowed residue limit) under the Federal Food, Drug, and Cosmetic Act.

4.2.3. Air

The Clean Air Act (CAA) governs, among other things, the establishment, review and revision of national ambient air quality standards and identification of criteria air pollutants. As amended in 1990, it also identified 190 Hazardous Air Pollutants (HAPs) for regulation (the list currently includes 187 HAPs) and provides EPA with authority to identify additional HAPs. The CAA also contains requirements that address accidental releases of hazardous substances from stationary sources that potentially can have serious adverse effects to human health or the environment. Use or manufacture of nanomaterials could result in emissions of pollutants that are or possibly could be listed as criteria air pollutants or HAPs.

Under the CAA, EPA has issued health effects testing requirements for fuels and fuel additives. Gasoline and diesel fuels and their additives are subject to the regulations issued by EPA. These fuels and additives for use in on-road applications may not be introduced into commerce until they have been registered by EPA. As previously noted EPA has received and is reviewing an application for registration of a diesel additive containing cerium oxide.

4.2.4. Pollution Prevention

The Pollution Prevention Act of 1990 was considered a turning point in how the nation looks at the control of pollution. Instead of focusing on waste management and pollution control, Congress declared a national policy for the United States to address pollution based on "source reduction." The policy established a hierarchy of measures to protect human health and the environment, where multi-media approaches would be anticipated: (1) pollution should be prevented or reduced at the source; (2) pollution that cannot be prevented should be recycled in an environmentally safe manner; (3) pollution that cannot be prevented or recycled should be treated in an environmentally safe manner; and (4) disposal or other release into the environment should be employed only as a last resort and should be conducted in an environmentally safe manner.

As a result of the Act, two programs were initiated, with two different approaches, to meet the spirit of the new national policy: the Design for the Environment (DfE) Program and the Green Chemistry Program. Under DfE, EPA works in partnership with industry sectors to improve performance of commercial processes while reducing risks to human health and the environment. The Green Chemistry Program promotes research to design chemical products and processes that reduce or eliminate the use and generation of toxic chemical substances. In 1998, EPA complimented these two programs with the Green Engineering Program, which applies approaches and tools for evaluating and reducing the environmental impacts of processes and products (see http://www.epa.gov/oppt/greenengineering). Nanotechnology offers an opportunity to implement pollution prevention principles into the design of a new technology.

4.2.5. Water

The stated goals of the Clean Water Act (CWA) are to protect the chemical, physical, and biological integrity of the nation's waters as well as to ensure the health and welfare of the environment, fish, shellfish, other aquatic organisms, wildlife, and humans that live in, recreate on, or come in contact with waters of the United States. Depending on the toxicity of nanomaterials to aquatic life, aquatic dependent wildlife, and human health, as well as the potential for exposure, nanomaterials may be regulated under the CWA. A variety of approaches are available under the CWA to provide protection, including effluent limitation guidelines, water quality standards (aquatic life, human health, biological), best management practices, NPDES permits, and whole effluent toxicity testing. Simultaneously, nanomaterials may provide an effective and efficient mechanism to resolve water quality contamination and its impacts on aquatic life, aquatic dependent wildlife, and human health. Both scenarios must be explored to determine how and when to regulate these potentially hazardous additions to the nation's waters.

The Safe Drinking Water Act (SDWA), as amended in 1996, is the main federal law that protects public health by regulating hazardous contaminants in drinking water. SDWA authorizes the Agency to establish non-enforceable health-based Maximum Contaminant Level Goals (MCLGs) and enforceable Maximum Contaminant Levels (MCLs) or required treatment techniques, as close as feasible to the MCLGs, taking into consideration costs and available analytical and treatment technology. Nanotechnology has the potential to influence the setting of MCLs through improvements in analytical methodology or treatment techniques. Nanotechnology has the potential to contribute to better and more cost-effective removal of drinking water contaminants, such as metals (e.g. arsenic or chromium), toxic halogenated organic chemicals, suspended particulate matter and pathogenic microorganisms. If nanoparticles enter drinking water, such as through their use in water treatment, then exposure to nanomaterials may occur through drinking water ingestion or inhalation (e.g. from showering). Based on their toxicity and occurrence in drinking water supplies, nanomaterials could be regulated under the SDWA.

4.2.6. Solid Waste

The Comprehensive Environmental Response, Compensation, and Liability Act (CERCLA) addresses contamination at closed and abandoned waste sites. CERCLA gives EPA the authority to respond to actual or threatened releases of hazardous substances to the environment and to actual or threatened releases of pollutants or contaminants that may present an imminent and substantial danger to the public health or welfare. Nanomaterials that meet these criteria potentially would be subject to this authority.

The Resource Conservation and Recovery Act (RCRA), which amended the Solid Waste Disposal Act, regulates, from the point of generation, the management of solid and hazardous wastes, underground storage tanks, and medical wastes. Subtitle D of RCRA covers municipal and other non-hazardous wastes. Subtitle C of RCRA covers the storage, transportation, treatment, disposal, and cleanup of hazardous wastes. Nanomaterials that meet one or more of the definitions of a hazardous waste (i.e., a waste that is specifically listed in the regulations and/or that exhibits a defining characteristic) potentially would be subject to subtitle C regulations. Subtitle I covers underground storage tanks, and Subtitle J covers medical waste incineration.

The 1990 Oil Pollution Act (OPA) amended the Clean Water Act (CWA) to address the harmful environmental impacts of oil spills. EPA responsibilities under the Oil Pollution Act include response (cleanup/containment/prevention action) and enforcement actions related to discharges and threatened discharges of oil or hazardous substances in the inland waters of the United States.

4.2.7. Toxics Release Inventory Program

In 1986, Congress passed the Emergency Planning and Community Right to Know Act (EPCRA) and the Toxics Release Inventory (TRI) was established. The TRI is a publicly available database containing information on toxic chemical releases and other waste management activities that are reported annually by manufacturing facilities and facilities in certain other sectors, as well as federal facilities. Some producers of nanomaterials containing materials listed in the TRI may be subject to reporting under the TRI Program (www.epa.gov/tri/). Facilities required to report TRI chemical releases and other waste management quantities are those that met or exceeded the minimum criteria of number of employees and total mass of chemical manufactured, processed, or otherwise used in a calendar year. Of the nearly 650 toxic chemicals and chemical compounds on the TRI, a number are metals and compounds containing these metals, including cadmium, chromium, copper, cobalt and antimony. Such compounds may be produced as nanomaterials, and some are commonly used in quantum dots.

4.3. ENVIRONMENTAL STEWARDSHIP

Nanotechnology provides an opportunity for EPA and other stakeholders to develop best practices for preventing pollution at its source and conserving natural resources whenever possible. For example, EPA and others are supporting research into *green nanotechnology*, to identify applications of nanotechnology that reduce pollution from industrial processes as well as to develop manufacturing process that fabricate nanomaterials in an environmentally friendly manner. Appendix B provides a fuller discussion of stewardship principles. Many diverse industrial organizations and their suppliers have the opportunity at this early stage of technology development and use to be leading environmental stewards.

At EPA, in addition to our support for green nanotechnology research, there are a number of programs already in place that are based upon environmental stewardship principles. These programs address processes, including inputs; waste streams; and the design, use, disposal, and stewardship of products consistent with the goal of pollution prevention. Information on nanotechnologies and materials could be provided through existing information networks, and EPA could pursue additional voluntary initiatives or integrate nanotechnology and nanoscale materials into already existing voluntary programs to ensure responsible development.

In: Nanotechnology and the Environment
Editor: Robert V. Neumann

ISBN: 978-1-60692-663-5
© 2010 Nova Science Publishers, Inc.

Chapter 5

EPA'S RESEARCH NEEDS FOR NANOMATERIALS

U.S. Environmental Protection Agency

Research is needed to inform EPA's actions related to the benefits and impacts of nanomaterials. However, there are significant challenges to addressing research needs for nanotechnology and the environment. The sheer variety of nanomaterials and nanoproducts adds to the difficulty of developing research needs. Each stage in their lifecycle, from extraction to manufacturing to use and then to ultimate disposal, will present separate research challenges. Nanomaterials also present a particular research challenge over their macro forms in that we have a very limited understanding of nanoparticles' physicochemical properties. Research should be designed from the beginning to identify beneficial applications and to inform risk assessment, pollution prevention, and potential risk management methods. Such research will come from many sources, including academia, industry, EPA, and other agencies and organizations. Other government and international initiatives have also undertaken efforts to identify research needs for nanomaterials and have come to similar conclusions (UK Department for Environment, Food and Rural Affairs, 2005; NNI, 2006c).

An overarching, guiding principle for all testing, both human health and ecological, is the determination of which nanomaterials are most used and/or have potential to be released to, and interact with, the environment. These nanomaterials should be selected from each of the broader classes of nanomaterials (carbon-based, metal-based, dendrimers, or composites) to serve as representative particles for testing/evaluation purposes.

5.1. RESEARCH NEEDS FOR ENVIRONMENTAL APPLICATIONS

The Agency recognizes the benefits of using nanomaterials in environmental technologies. Research is needed to develop and test the efficacy of applications that detect, prevent and clean up contaminants. EPA also has the responsibility for determining the ecological and human health implications of these technologies.

5.1.1. Green Manufacturing Research Needs

Nanotechnology offers the possibility of changing manufacturing processes in at least two ways: (1) by using less materials and (2) using nanomaterials for catalysts and separations to increase efficiency in current manufacturing processes. Nanomaterial and nanoproduct manufacturing offers the opportunity to employ the principles of green chemistry and engineering to prevent pollution from currently known harmful chemicals. Research enabling this bottom-up manufacturing of chemicals and materials is one of the most important areas in pollution prevention in the long term. Research questions regarding green manufacturing include:

- How can nanotechnology be used to reduce waste products during manufacturing?
- How can nanomaterials be made using benign starting materials?
- How can nanotechnology be used to reduce the resources needed for manufacturing (both materials and energy)?
- What is the life cycle of various types of nanomaterials and nanoproducts under a variety of manufacturing and environmental conditions?

5.1.2. Green Energy Research Needs

Developing green energy approaches will involve research in many areas, including solar energy, hydrogen, power transmission, diesel, pollution control devices, and lighting. These areas have either direct or indirect impacts on environmental protection. In solar energy, nanomaterials may make solar cells more efficient and more affordable. In addition, nanocatalysts may efficiently create hydrogen from water using solar energy. Research questions for green energy include:

- What research is needed for incentives to encourage nanotechnology to enable green energy?
- How can nanotechnology assist "green" energy production, distribution, and use?

5.1.3. Environmental Remediation/Treatment Research Needs

The research questions in this area revolve around the effectiveness and risk parameters of nanomaterials to be used in site remediation. Materials such as zero-valent iron are expected to be useful in replacing current pump-and-treat or off site treatment methods. In addition, other nanoremediation approaches can involve the methods of coating biological particles, determining the effect on the particles (enzyme or bacteriophage) following coating, and application technologies. This is an area that has not been examined in any great detail. Therefore, research is needed to develop technologies using nanocoated biological particles for environmental decontamination or prophylactic treatment to prevent contamination. The products of this research would be technologies utilizing innocuous biological entities treated with nanoparticles to decontaminate or prevent bacterial growth. In an age of antibiotic resistance and aversion to chemical decontamination, enzyme and bacteriophage technologies offer an attractive option. Remediation and treatment research questions include:

- Which nanomaterials are most effective for remediation and treatment?
- What are the fate and effects of nanomaterials used in remediation applications? When nanomaterials are placed in groundwater treatment, how do they behave over time? Do they move in groundwater? What is their potential for migrating to drinking water wells?
- How can we improve methods for detecting and monitoring nanomaterials used in remediation and treatment?
- To what extent are these materials and their byproducts persistent, bioaccumulative, and toxic and what organisms are affected?
- If toxic byproducts are produced, how can these be reduced?
- What is needed to enhance the efficiency and cost-effectiveness of remediation and treatment technology?

5.1.4. Sensors

In general, nanosensors can be classified in two main categories: (1) sensors that are used to measure nanoscale properties (this category comprises most of the current market) and (2) sensors that are themselves nanoscale or have nanoscale components. The second category can eventually result in lower material cost as well as reduced weight and power consumption of sensors, leading to greater applicability, and is the subject of this section. Research needs for sensors to detect nanomaterials in the environment are discussed in the Environmental Detection section below.

- How can nanomaterials be employed in the development of sensors to detect biological and chemical contaminants?
- How can sensor systems be developed to monitor agents in real time and the resulting data accessed remotely?
- How these small-scale monitoring systems be developed to detect personal exposures and *in vivo* distributions of toxicants.

5.2. RESEARCH NEEDS FOR RISK ASSESSMENT

5.2.1. Chemical Identification and Characterization

Research that can be replicated depends on agreement on the identification and characterization of nanomaterials. In addition, understanding the physical and chemical properties in particular is necessary in the evaluation of hazard (both human and ecological) and exposure (all routes). It is not clear whether existing physical-chemical property test methods are adequate for sufficiently characterizing various nanomaterials. Alternative methods may be needed. Research questions include:

- What are the unique chemical and physical characteristics of nanomaterials? How do these characteristics vary among different classes of materials (e.g., carbon based, metal based) and among the individual members of a class (e.g., fullerenes, nanotubes)?
- How do these properties affect the material's reactivity, toxicity and other attributes?
- To what extent will it be necessary to tailor research protocols to the specific type and use pattern of each nanomaterial? Can properties and effects be extrapolated within a class of nanomaterials?
- Are there adequate measurement methods/technology available to fully characterize nanomaterials, to distinguish among different types of nanomaterials, and to distinguish intentionally produced nanomaterials from ultrafine particles or naturally occurring nanosized particles?
- Are current test methods for characterizing nanomaterials adequate for the evaluation hazard and exposure data?
- Do nanomaterial characteristics vary from their pure form in the laboratory to their form as components of products and eventually to the form in which they occur in the environment?
- What intentionally produced nanomaterials are now on the market and what new types of materials can be expected to be developed?
- How will manufacturing processes, formulations, and incorporations in end products alter the characteristics of nanomaterials?

5.2.2. Environmental Fate and Treatment Research Needs

EPA needs to ascertain the fate of nanomaterials in the environment to understand the availability of these materials for exposure to humans and other life forms. Research on the transport and potential transformation of nanomaterials in soil, subsurface, surface waters, wastewater, drinking water, and the atmosphere is essential as nanomaterials are used increasingly in products. To support these investigations, existing methods should be evaluated and if necessary, they should be modified or new methods should be developed. Research is needed to address the following high-priority questions.

5.2.2.1. Transport Research Questions

- What are the physicochemical factors that influence the transport and deposition of intentionally produced nanomaterials in the environment? How do nanomaterials move through these media? Can existing information on soil colloidal fate and transport and atmospheric ultrafine particulate fate and transport inform our thinking?
- How are nanomaterials transported in the atmosphere? What nanomaterial properties and atmospheric conditions control the atmospheric fate of nanomaterials?
- To what extent are nanomaterials mobile in soils and in groundwater? What is the potential for these materials, if released to soil or landfills, to migrate to groundwater

and within aquifers, with potential exposure to general populations via groundwater ingestion?

- What is the potential for these materials to be transported bound to particulate matter, sediments, or sludge in surface waters?
- How do the aggregation, sorption and agglomeration of nanoparticles affect their transport?
- How do nanomaterials bioaccumulate? Do their unique characteristics affect their bioavailability? Do nanomaterials bioaccumulate to a greater or lesser extent than macro-scale or bulk materials?

5.2.2.2. Transformation Research Questions

- How do nanoparticles react differently in the environment than their bulk counterparts?
- What are the physicochemical factors that affect the persistence of intentionally produced nanomaterials in the environment? What data are available on the physicochemical factors that affect the persistence of unintentionally produced nanomaterials (e.g., carbon-based combustion products) that may provide information regarding intentionally produced nanomaterials?
- Do particular nanomaterials persist in the environment, or undergo degradation via biotic or abiotic processes? If they degrade, what are the byproducts and their characteristics? Is the nanomaterial likely to be in the environment, and thus be available for bioaccumulation/biomagnification?
- How are the physicochemical and biological properties of nanomaterials altered in complex environmental media such as air, water, and soil? How do redox processes influence environmental transformation of nanomaterials? To what extent are nanomaterials photoreactive in the atmosphere, in water, or on environmental surfaces?
- How do the aggregation, sorption and agglomeration of nanoparticles affect transformation?
- In what amounts and in what forms may nanoparticles be released from materials that contain them, as a result of environmental forces (rain, sunlight, etc.) or through use, re-use, and disposal.

5.2.2.3. Chemical Interaction Research Questions

- How do nanosized adsorbants and chemicals sorbed to them influence their respective environmental interactions? Can these materials alter the mobility of other substances in the environment? Can these materials alter the reactivity of other substances in the environment?

5.2.2.4. Treatment Research Questions

- What is the potential for these materials to bind to soil, subsurface materials, sediment or wastewater sludge, or binding agents in waste treatment facilities?

- Are these materials effectively removed from wastewater using conventional wastewater treatment methods and, if so, by what mechanism?
- Do these materials have an impact on the treatability of other substances in waste streams (e.g., wastewater, hazardous and nonhazardous solid wastes), or on treatment facilities performance?
- Are these materials effectively removed in drinking water treatment and, if so, by what mechanism?
- Do these materials have an impact on the removal of other substances during drinking water treatment, or on drinking water treatment facilities performance?
- How effective are existing treatment methods (e.g., carbon adsorption, filtration, coagulation and settling, or incineration/air pollution control system sequestration/stabilization) for treating nanomaterials?

5.2.2.5. *Assessment Approaches and Tools Questions*

- Can existing information on soil colloidal fate and transport, as well as atmospheric ultrafine particulate fate and transport, inform our thinking? Do the current databases of ultrafines/fibers shed light on any of these questions?
- Do the different nanomaterials act similarly enough to be able to create classes of like compounds? Can these classes be used to predict structure-activity relationships for future materials?
- Should current fate and transport models be modified to incorporate the unique characteristics of nanomaterials?

5.2.3. Environmental Detection and Analysis Research Needs

While there are a variety of methods currently available to measure nanoparticle mass/mass concentrations, surface area, particle count, size, physical structure (morphology), and chemical composition in the laboratory, the challenge remains to detect nanomaterials in the environment. Research is needed to address the following high-priority questions:

5.2.3.1. *Existing Methods and Technologies Research Questions*

- Are existing methods and technologies capable of detecting, characterizing, and quantifying intentionally produced nanomaterials by measuring particle number, size, shape, surface properties (e.g., reactivity, charge, and area), etc.? Can they distinguish between intentionally produced nanomaterials of interest and other ultrafine particles? Can they distinguish between individual particles of interest and particles that may have agglomerated or attached to larger particles?
- Are standard procedures available for both sample preparation and analysis?
- Are quality assurance and control reference materials and procedures available?
- How would nanomaterials in waste media be measured and evaluated?

5.2.3.2. New Methods and Technologies Research Needs

- What low-cost, portable, and easy-to-use technologies can detect, characterize, and quantify nanomaterials of interest in environmental media and for personal exposure monitoring.

5.2.4. Human Exposures, Their Measurement and Control

Potential sources of human exposure to nanomaterials include workers exposed during the production and use of nanomaterials, general population exposure from releases to the environment during the production or use in the workplace, and direct general population exposure during the use of commercially available products containing nanoscale materials. Releases from industrial accidents, natural disasters, or malevolent activity such as a terrorist attack should also be considered. Research is needed to identify potential sources, pathways, and routes of exposure, potential tools and models that may be used to estimate exposures, and potential data sources for these models, as well as approaches for measuring and mitigating exposure. NIOSH has also examined research needs regarding risks to workers and developed a strategic plan to address these needs (NIOSH 2005a, b). Research is needed to address the following high-priority questions.

5.2.4.1. Risk and Exposure Assessment Research Questions

- Is the current exposure assessment process adequate for assessing exposures to nanomaterials? Is mass dose an effective metric for measuring exposure? What alternative metric (e.g., particle count, surface area) should be used to measure exposure? Are sensitive populations' (e.g., endangered species, children, asthmatics, etc.) exposure patterns included?
- How do physical and chemical properties of nanomaterials affect releases and exposures?
- How do variations in manufacturing and subsequent processing, and the use of particle surface modifications affect exposure characteristics?

5.2.4.2. Release and Exposure Quantification Research Questions

- What information is available about unique release and exposure patterns of nanomaterials? What additional information is needed?
- What tools/resources currently exist for assessing releases and exposures within EPA (chemical release information/monitoring systems (e.g., TRI), measurement tools, models, etc)? Are these tools/resources adequate to measure, estimate, and assess releases and exposures to nanomaterials? Is degradation of nanomaterials accounted for?
- What research is needed to develop sensors that can detect nanomaterials, including personal exposure monitoring?

5.2.4.3. Release and Exposure Reduction and Mitigation Research Questions

- What tools/resources exist for limiting release and/or exposure during manufacture, use or following release via waste streams? Are these tools/resources adequate for nanomaterials?
- Are current respirators, filters, gloves, and other PPE capable of reducing or eliminating exposure from nanomaterials?
- Are current engineering controls and pollution prevention devices capable of minimizing releases and exposures to nanomaterials?
- Are technologies and procedures for controlling spills during manufacture and use adequate for nanomaterials? Can current conventional technologies (i.e., for non-nanomaterials) be adapted to control nanomaterial spills?
- In the case of an unintentional spill, what are the appropriate emergency actions? How are wastes from response actions disposed of properly?
- Do existing methods using vacuum cleaners with HEPA filters work to clean up a spill of solid nanomaterials? If not, would a wet vacuum system work?
- What PPEs would be suitable for use by operators during spill mitigation?

5.2.5. Human Health Effects Assessment Research Needs

Adverse health effects of intentionally produced nanomaterials may result from either direct exposure resulting from inadvertent release of these novel materials or unintentional byproducts produced by their intentional release into the environment. Very little data exist on the toxicity, hazardous properties, deposition and fate, as well as susceptibility associated with exposure to intentionally produced nanomaterials, their application byproducts, decomposition products or production waste streams. Finally, it is uncertain whether standard test methods will be capable of identifying toxicities associated with the unique physical chemical properties of intentionally produced nanomaterials.

It will be important for nanomaterial health effects risk assessment research to also establish whether current particle and fiber toxicological databases have the ability to predict or assess the toxicity of intentionally produced nanomaterials displaying unique physicochemical properties. The limited studies conducted to date indicate that the toxicological assessment of specific intentionally produced nanomaterials will be difficult to extrapolate from existing databases. The toxic effects of nanoscale materials have not been fully characterized, but it is generally believed that nanoparticles can have toxicological properties that differ from their bulk material. A number of studies have demonstrated that nanoparticle toxicity is complex and multifactorial, potentially being regulated by a variety of physiochemical properties such as size, chemical composition, and shape, as well as surface properties such as charge, area and reactivity. As the size of particles decreases, a resulting larger surface-to-volume ratio per unit weight for nanoparticles correlates with increased toxicity as compared with bulk material toxicity. Also as a result of their smaller size, nanoparticles may pass into cells directly through cell membranes or penetrate the skin and distribute throughout the body once translocated to the circulatory system. While the effects of shape on toxicity of nanoparticles appears unclear, the results of a recent *in vitro* cytotoxicity study appear to suggest that single-wall carbon nanotubes are more toxic than multi-wall carbon nanotubes. Therefore, with respect to nanoparticles, there is concern for systemic effects (e.g., target organs, cardiovascular, and neurological toxicities) in addition to portal-of-entry (e.g. lung, skin, intestine) toxicity.

Initially, it will be important to be specific with respect to the nature of the surface material/coating, the application for which the material is used, the likely route of exposure, the presence of other exposures which may affect toxicity (e.g., UV radiation) and not rely on information derived from a study conducted under one set of conditions to predict outcomes that may occur under another set of conditions. However, past experience with conventional chemicals suggests that toxicology research on nanomaterials should be designed from the beginning with an eye towards developing hypothesis-based predictive testing.

Research is also needed to examine health impacts of highly dispersive nanotechnologies that are employed for site remediation, monitoring, and pollution control strategies. It will be necessary to determine both the impacts these types of nanotechnologies have on regulated pollutants in air, soil, or water, as well as their corresponding potential health effects. Research should be conducted in the following areas:

A Determining the adequacy of current testing schemes, hazard protocols, and dose metrics.
B Identifying the properties of nanomaterials that are most predictive of toxicity to receptors and their sensitive subpopulations.
C Identifying those nanomaterials with high commercial potential with dispersive applications, and their most probable exposure pathways.

These areas lead to the following research questions:

- What are the health effects (local and systemic; acute and chronic) from either direct exposure to nanomaterials, or to their byproducts, associated with those nanotechnology applications that are most likely to have potential for exposure? (Addresses area C, above)
- Are there specific toxicological endpoints that are of higher concern for nanomaterials, such as neurological, cardiovascular, respiratory, or immunological effects, etc.? (Addresses area C, above)
- Are current testing methods (organisms, exposure regimes, media, analytical methods, testing schemes) applicable to testing nanomaterials in standardized agency toxicity tests (http://www.epa.gov/opptsfrs/OPPTS_Harmonized/)? (Addresses area A, above)
- Are current test methods, for example OECD and EPA harmonized test guidelines, capable of determining the toxicity of the wide variety of intentionally produced nanomaterials and byproducts associated with their production and applications? (Addresses area A, above)
- Are current analytical methods capable of analyzing and quantifying intentionally produced nanomaterials to generate dose-response relationships? (Addresses area A, above)

- What physicochemical properties regulate nanomaterial absorption, distribution, metabolism, and excretion (ADME)? (Addresses area A, above)
- What physicochemical properties and dose metrics best correlate with the toxicity (local and systemic; acute and chronic) of intentionally produced nanomaterials following various routes of exposure? (Addresses area A, above)
- How do variations in manufacturing and subsequent processing, and the use of particle surface modifications affect nanomaterial hazard? (Addresses area B, above)
- Are there subpopulations that may be at increased risk of adverse health effects associated with exposure to intentionally produced nanomaterials? (Addresses area B, above)
- What are the best approaches to build effective predictive models of toxicity (SAR, PBPK, "omics", etc.)? (Addresses areas A and B, above)
- Are there approaches to grouping particles into classes relative to their toxicity potencies, in a manner that links *in vitro, in vivo,* and *in silico* data?

5.2.6. Ecological Effects Research Needs

Ecosystems may be affected through inadvertent or intentional releases of intentionally produced nanomaterials. Drug and gene delivery systems, for example, are not likely to be used directly in the environment but may contaminate soils or surface waters through waste water treatment plants (from human use) or more directly as runoff from concentrated animal feeding operations (CAFOs) or from aquaculture. Direct applications may include nanoscale monitoring systems, control or clean-up systems for conventional pollutants, and desalination or other chemical modifications of soil or water. Nanoscale particles may affect aquatic or terrestrial organisms differently than larger particles due to their extreme hydrophobicity, their ability to cross and/or damage cell membranes, and differences in electrostatic charge. Furthermore, use of nanomaterials in the environment may result in novel byproducts or degradates that also may pose significant risks.

Important considerations for prioritizing and defining the scope of the research needs include determining which nanomaterials are most used (volume), are likely to be used in the near future (imminence of use), and/or have most potential to be released into the environment (dispersive applications). Another consideration is the need to test representative materials from each of the four classes of nanomaterials (carbon-based, metal-based, dendrimers, composites).

The same general research areas used for prioritizing human health effects research needs were used to prioritize ecological research needs. Using these areas as a guide, the following research questions were identified:

- Are current testing schemes and methods (organisms, endpoints, exposure regimes, media, analytical methods) applicable to testing nanomaterials in standardized toxicity tests? Both pilot testing protocols and definitive protocols should be evaluated with respect to their applicability to nanomaterials.
- What is the distribution of nanomaterials in ecosystems? Research on model ecosystems studies (micro, mesocosms) is needed to assist in determining the

distribution of nanomaterials in ecosystems and potentially affected compartments and species.

- What are the effects (local and systemic; acute and chronic) from either direct exposure to nanomaterials, or to their byproducts, associated with those nanotechnology applications that are most likely to have potential for exposure?
- What are the absorption, distribution, metabolism, elimination (ADME) parameters for various nanomaterials for ecological receptors? This topic addresses the uptake, transport, bioaccumulation relevant to a range of species (fish, invertebrates, birds, amphibians, reptiles, plants, microbes).
- How do variations in manufacturing and subsequent processing, and the use of particle surface modifications affect nanomaterial toxicity to ecological species?
- What research is needed to examine the interaction of nanomaterials with microbes in sewage treatment plants, in sewage effluent, and in natural communities of microbes in natural soil and natural water?
- What research is needed to develop structure activity relationships (SARs) for nanomaterials for aquatic organisms?
- What are the modes of action (MOAs) for various nanomaterials for ecological species? Are the MOAs different or similar across ecological species?

5.2.7. Risk Assessment Research - Case Study

The overall risk assessment approach used by EPA for conventional chemicals is thought to be generally applicable to nanomaterials. It will be necessary to consider nanomaterials' special properties and their potential impacts on fate, exposure, and toxicity in developing risk assessments for nanomaterials. It may be useful to consider a tiered-testing scheme in the development of testing and risk assessment approaches to nanomaterials. Also, decisions will need to be made even as preliminary data are being generated, meaning that decision making will occur in an environment of significant uncertainty. Decision-support tools will need to be developed and applied to inform assessments of potential hazard and exposure.

Case studies could be conducted based on publicly available information on several intentionally produced nanomaterials. Such case studies would be useful in further identifying unique considerations that should be focused in conducting risk assessments for various types of nanomaterials. From such case studies and other information, information gaps may be identified, which can then be used to map areas of research that are directly affiliated with the risk assessment process and the use of standard EPA tools such as tiered testing schemes. EPA frequently uses tiered testing schemes for specific risk assessment applications. A series of workshops involving a substantial number of experts from relevant disciplines could be held to use case studies and other information for the identification of any unique considerations for nanomaterials not previously identified, testing schemes, and associated research needs that will have to be met to carry out exposure, hazard and risk assessments.

In: Nanotechnology and the Environment
Editor: Robert V. Neumann

ISBN: 978-1-60692-663-5
© 2010 Nova Science Publishers, Inc.

Chapter 6

RECOMMENDATIONS

U.S. Environmental Protection Agency

This section provides staff recommendations for Agency actions related to nanotechnology. These staff recommendations are based on the discussion of nanotechnology environmental applications and implications discussed in this paper, and are presented to the Agency as proposals for EPA actions for science and regulatory policy, research and development, collaboration and communication, and other Agency initiatives. Included below are staff recommendations for research that EPA should conduct or otherwise fund to address the Agency's decision-making needs. When possible, relative priorities have been given to these needs. Clearly, the ability of EPA to address these research needs will depend on available resources and competing priorities. Potential lead offices in the Agency have been identified for each recommendation. It may be appropriate for other EPA offices to collaborate with the identified leads for specific recommendations. EPA should also collaborate with outside groups to avoid duplication and leverage research by others. Identified research recommendations were used as a point of departure for Agency discussion and development of the EPA Nanotechnology Research Framework, attached as Appendix C.

6.1. RESEARCH RECOMMENDATIONS FOR ENVIRONMENTAL APPLICATIONS

6.1.1. Research Recommendations for Green Manufacturing

- ORD and OPPT should take the lead in investigating and promoting ways to apply nanotechnology to reduce waste products generated, and energy used, during manufacturing of conventional materials as well as nanomaterials.

6.1.2. Research Recommendations for Green Energy

- ORD and OPPT should promote research into applications of nanomaterials green energy approaches, including solar energy, hydrogen, power transmission, diesel, pollution control devices, and lighting.

6.1.3. Environmental Remediation/Treatment Research Needs

- ORD should support research on improving pollutant capture or destruction by exploiting novel nanoscale structure-property relations for nanomaterials used in environmental control and remediation applications.

6.1.4. Research Needs for Sensors

- ORD should support development of nanotechnology-enabled devices for measuring and monitoring contaminants and other compounds of interest, including nanomaterials. For example, ORD should lead development of new nanoscale sensors for the rapid detection of virulent bacteria, viruses, and protozoa in aquatic environments

6.1.5. Research Needs for Other Environmental Applications

- ORD should work with industrial partners to verify the performance of nanomaterials and nanoproducts used for environmental applications.
- ORD should develop rapid screening methods that keep pace with rapid technological change for nanomaterials and nanoproducts building on existing Life Cycle Analysis methods. OPPTS, OW and OAR should collaborate with stakeholders developing rapid screening methods.
- ORD and OPPT should collaborate with NIOSH and others to evaluate the application of nanotechnology for exposure reduction; e.g., nano-enabled PPE, respirators containing nanomaterials, and nanoscale end-of-life sensors, sensors that indicate when a product has reached the end of its useful life.

6.2. RESEARCH RECOMMENDATIONS FOR RISK ASSESSMENT

A multidisciplinary approach is needed that involves physics, biology, and chemistry to understand nanomaterials at a basic level and how they interact with the environment. This calls for a cross-media approach and one that involves collaboration with other federal agencies, and the private and non-profit sectors. This includes examining the implications (risks) of the environmental applications of nanotechnology.

6.2.1 Research Recommendations for Chemical Identification and Characterization

- ORD should lead research on the unique chemical and physical characteristics of nanomaterials and how these properties affect the material's reactivity, toxicity and other attributes.
- ORD should lead research on how nanomaterial characteristics vary from their pure form in the laboratory to their form as components of products, and eventually to the form in which they occur in the environment.
- ORD should determine if there are adequate measurement methods/technology available to fully characterize nanomaterials, to distinguish among different types of nanomaterials, and to distinguish intentionally produced nanomaterials from ultrafine particles or naturally occurring nanosized particles.

6.2.2. Research Recommendations for Environmental Fate and Treatment

The following are recommendations, in order of priority, in support of the environmental fate and treatment research needs identified as priorities in Chapter 5.

Fate, Treatment and Transport

- OSWER and ORD should lead research on the fate of nanomaterials, such as zero-valent iron, used in the remediation of chemically contaminated sites. This research should also address the impacts of such nanomaterials on the fate of other contaminants at remediation sites. These offices should collaborate with state environmental programs and academia on this research. Based upon available field activities where nanomaterials are being used for site remediation, this research could be conducted within the few years.
- ORD and OAR should lead research on the stability of various types of intentionally produced nanoparticles in the atmosphere. This effort should involve both theoretically derived information as well as some laboratory testing.
- ORD, OPPT, OPP, OSWER and OW should lead research on the biotic and abiotic transport and degradation of nanomaterials waters, soils and sediment that are relevant to environmental conditions.
- ORD should lead research that defines the physical and chemical properties of nanomaterials that impact their environmental fate.
- ORD, OSW and OW should collaboratively lead research on treatment methods used for removing nanomaterials from wastewater. Research should include analysis of the specific types of nanomaterials that are likely to end up in large quantities in sewage treatment plants, the efficiency of removing nanoparticles from the effluent, the fate of nanomaterials after removal, methods for disposal of sludges containing nanomaterials, and the impact nanomaterials may have on the removal or degradation of other substances in sewage during the treatment process. EPA should collaborate with municipal sewage treatment facilities and academia on this research.

- ORD, OPPT and OW should share the lead on research on the fate of nanomaterials used in the purification of drinking water. Research would be based on actual field and/or laboratory findings and recommendations would be provided on how to improve the nanomaterial removal process where human health issues are a concern. This research should also evaluate individual processes; i.e., whether methods such as carbon adsorption, filtration, and coagulation and settling are effective for treating nanomaterials.

- ORD, OSW and OAR should lead research on the fate of nanomaterials in incineration and other thermal treatment processes, including the efficiency of destroying nanomaterials, the efficiency of various air pollution control devices (e.g., baghouses, liquid scrubbers, and electrostatic precipitators) at removing entrained nanomaterials, the fate of nanomaterials after removal, methods for disposal of ash containing nanomaterials, and the impact nanomaterials may have on the removal or degradation of other substances during the treatment process.

- ORD and OSW should lead research on the fate of nanomaterials in other waste treatment processes (e.g. chemical oxidation, stabilization). Research would identify relevant waste streams, the efficiency of current treatment regimes at addressing nanomaterials, the fate of nanomaterials after treatment, methods for disposal of treatment output containing nanomaterials, and the impact nanomaterials may have on the treatment of other toxic constituents in the waste stream. EPA should collaborate with treatment, storage, and disposal facilities (TSDFs) and academia on this research.

- ORD and OSW should lead research on the fate of nanomaterials in municipal, industrial, and hazardous waste (i.e., Subtitle C) landfills, and other land-based waste management scenarios (e.g., surface impoundments). Research would identify relevant waste streams, the efficiency of current containment technologies (e.g., various cap and liner types, leachate collection systems) at preventing the leaching of nanomaterials into groundwater, the fate of nanomaterials after disposal, and the impact nanomaterials may have on the containment of other toxic constituents in the waste stream. EPA should collaborate with municipal and industrial stakeholders, and academia on this research.

6.2.3. Research Recommendations for Environmental Detection and Analysis

Where applicable, the initial focus of environmental detection and analysis related research should be on nanomaterials or types of nanomaterials that have demonstrated potential human or ecological toxicity. The following is a prioritized list of research needs for environmental detection and analysis.

- ORD should lead the development of a report on the assessment of available environmental detection methods and technologies for nanomaterials in environmental media and for personal exposure monitoring. ORD could collaborate with NIOSH, DOD, industry and academia in developing this chapter.

- ORD should collaborate with NIST, NIOSH, DOD, nanomaterial manufacturers and government and private sector organizations in the development of quality control reference materials for analytical methods for nanomaterials.
- ORD should lead development of a set of standard methods for the sampling and analysis for nanomaterials of interest in various environmental media. ORD should collaborate with NIOSH, DOD, industry, academia, the American Society for Testing Materials (ASTM) and the American National Standards Institute (ANSI) in developing these methods.

6.2.4. Research Recommendation Human Exposures, their Measurement and Control

The following is a prioritized list of research needs for human exposures, their measurement and control.

- OPPT should conduct a literature search to evaluate the effects of nanomaterial physical/chemical properties on releases and exposures.
- ORD and OPPT should lead research to determine what dose metrics (e.g. mass, surface area, particle count, etc.) are appropriate for measuring exposure to nanomaterials.
- OPPT and ORD should evaluate sources of information for assessing chemical releases and exposures for their applicability to nanomaterials. These sources, including release and exposure tools and models, would be evaluated to determine whether they would be applicable to assessing releases and exposures to nanomaterials. If found applicable, the sources would be evaluated to determine whether additional data or methods would be needed for assessing nanomaterials. Issues such as degradation would be considered also.
- OSWER, ORD, and OPPT should evaluate the proper emergency response actions and remediation in case of a nanomaterial spill, and the proper disposal of wastes from such response actions.
- OPPT should define risk assessment needs for various types of nanomaterials in consultation with other stakeholders.
- OPPT should consider approaches for performing exposure assessments for nanomaterials for human and environmental species, including sensitive populations (e.g., endangered species, children, asthmatics, etc.), in consultation with other offices and stakeholders.

Some parts of the remaining exposure and release research initiatives below are contingent upon completion of the risk and exposure assessment guidance documents noted in the two paragraphs above. Until this contingency is met, many of the remaining research needs cannot be fully completed.

- OPPT should lead development of exposure and release scenarios for nanomaterials in manufacturing and use operations. This effort should be conducted by OPPT in consultation with industry, NIOSH, and ORD, as appropriate.
- OPPT and ORD should evaluate and test equipment for controlling and reducing chemical releases and exposures for their applicability to nanomaterials.
- OPPT, ORD, OSWER, and OPP should evaluate and test personal protective equipment for controlling and reducing chemical exposures for their applicability to nanomaterials, in collaboration with NIOSH and other external groups.
- ORD should lead development of sensors for monitoring personal exposures to nanoparticles

6.2.5. Research Recommendations for Human Health Effects Assessment

The following is a prioritized list of health effects research needs:

Test Methods

- ORD and OPPTS should determine the applicability of current testing methods (organisms, exposure regimes, media, analytical methods, testing schemes) (http://www.epa.gov/opptsfrs/home/testmeth.htm) for evaluating nanoparticles in standardized agency toxicity tests. These offices should consider whether OECD and EPA harmonized test guidelines are capable of determining the toxicity of the wide variety of intentionally produced nanomaterials and waste byproducts associated with their production. In this effort ORD should lead research evaluating whether current analytical methods are capable of analyzing and quantifying intentionally produced nanomaterials to generate dose-response relationships.

Nanotoxicology

- ORD should lead research to determine the health effects (local and systemic; acute and chronic) resulting from either direct exposure to nanomaterials, or to their byproducts, associated with dispersive nanotechnology applications such as remediation, pesticides, and air pollution control technologies. Research should determine whether there are specific toxicological endpoints that are of high concern for nanoparticles, such as neurological, cardiovascular, respiratory, or immunological effects, etc. Research in this area should also provide information as to the adequacy of existing toxicological databases to predict or extrapolate the toxicity of intentionally produced nanomaterials. The Agency should also collaborate with stakeholders in catalyzing this research.

Hazard Identification and Dosimetry & Fate

- ORD should lead research to determine what physicochemical properties and dose metrics (mass, surface area, particle number, etc.) best correlate with the toxicity (local and systemic; acute and chronic) of intentionally produced nanomaterials.
- ORD should lead research on the absorption, distribution, metabolism, and excretion (ADME) of intentionally produced nanomaterials following various routes of exposure. This research must also include determining what physicochemical properties regulate intentionally produced nanomaterial ADME. ORD should collaborate with OPPTS on this research.

Susceptibility

- ORD should lead research to identify subpopulations that may be at increased risk for adverse health effects associated with exposure to intentionally produced nanomaterials. This is a need that cannot be established until information from earlier research needs have been collected.

Computational Nanotoxicology

- ORD should lead research to determine what approaches are most effective to build predictive toxicity assessment models (SAR, PBPK, "omics", etc.).

Research into the human health effects assessment of intentionally produced nanomaterials will be extremely challenging and the ability to interact with other federal, international, academic, and private activities in this area would be most beneficial. A number of organizations are engaged in health effects research. Collaboration with NASA, NIOSH, FDA, NCI, NTP, DOD/MURI, NIST, NEHI, DOE, the European Union, EPA grantees, academic institutions, and others will leverage resources in gaining knowledge on the potential health effects of nanomaterials.

6.2.6. Ecological Exposure and Effects

It is critical that research be focused specifically upon the fate, and subsequent exposure and effects, of nanomaterials on invertebrates, fish, and wildlife associated with ecosystems. What is the behavior of nano materials in aquatic and terrestrial environments? How can environmental exposures be simulated in the laboratory? What are the acute and chronic toxic effects? There is a need for development and validation of analytical methodologies for measuring nanoscale substances (both parent materials and metabolites/complexes) in environmental matrices, including tissues of organisms. In terms of toxicity, a critical challenge in the area of ecosystem effects lies in determining the impacts of materials whose cumulative toxicity is likely to be a manifestation of both physical and chemical interactions with biological systems. The following is a prioritized list of ecological research needs:

Test Methods

- ORD should collaborate with other EPA offices in research on the applicability of current testing schemes and methods (organisms, endpoints, exposure regimes, media, analytical methods) for testing nanomaterials in standardized toxicity tests. Both pilot testing protocols and definitive protocols should be evaluated with respect to their applicability to nanomaterials.

Environmental Fate/Distribution of Nanomaterials in Ecosystems

- ORD should lead on research on the distribution of nanomaterials in ecosystems.

Nanotoxicology and Dosimetry

- ORD should determine the effects of direct exposure to nanomaterials or their byproducts, associated with dispersive nanotechnology uses, on a range of ecological species (fish, inverts, birds, amphibians, reptiles, plants, microbes). This research should be focused on organisms residing in ecological compartments that the nanomaterials in question preferentially distribute to, if any, as identified above. This research should include evaluation of the uptake, transport, and bioaccumulation of these materials.
- ORD, OW and OPPT should lead research on the interactions of nanomaterials with microbes in sewage treatment plants in sewage effluent and natural communities of microbes in natural soil and natural water.
- ORD should lead research aimed at developing structure-activity relationships (SARs) for nanomaterials for aquatic organisms.
- ORD should lead research on the modes of action for various nanomaterials for a range of ecological species.

6.2.7. Recommendations to Address Overarching Risk Assessment Needs - Case Study

One way to examine how a nanomaterial assessment would fit within EPA's overall risk assessment paradigm is to conduct a case study based on publicly available information on one or several intentionally produced nanomaterials. In the past, such case studies have proven useful to the Agency in adjusting the chemical risk assessment process for stressors such as bacteria. For example, assessments of recombinant bacteria need to account for reproduction, and other factors not considered with chemical risk assessments. From such case studies and other information, information gaps may be identified, which can then be used to map areas of research that are directly affiliated with the risk assessment process. This has been done in the past with research on airborne particulate matter.

Additionally, a series of workshops involving a substantial number of experts from several disciplines should be held to use available information and principles in identifying data gaps and research needs that will have to be met to carry out exposure, hazard and risk assessments.

6.3. RECOMMENDATIONS FOR POLLUTION PREVENTION AND ENVIRONMENTAL STEWARDSHIP

Opportunities exist to advance pollution prevention as nanotechnology industries form and develop. EPA has the capability to support research into nanotechnology applications of pollution prevention and environmental stewardship principles that have been developed for green energy, green chemistry, green engineering, and environmentally benign manufacturing. EPA is well-positioned to work with stakeholders on pollution prevention and product stewardship approaches for producing nanomaterials in a green manner, as well as for identifying areas where nanomaterials may be cleaner alternatives to exisiting industrial inputs. The following are the primary recommendations for pollution prevention and environmental stewardship:

- EPA should support research into approaches that encourage environmental stewardship throughout the complete life cycle of nanomaterials and products.
- OPPT, ORD, and other stakeholders should encourage product stewardship, design for the environment, green engineering and green chemistry principles and approaches to nanomaterials and nanoproducts.

Other recommendations for pollution prevention and environmental stewardship:

- NCEI and OECA should research nanotechnology sectors, supply chains, analytical and design tools, and applications in order to inform pollution prevention approaches. OECA should collaborate with other Agency programs, such as OPPT's Green Supply Chain Network to identify nanotechnology sectors, supply chains, analytical and design tools, and applications.
- OCIR and OCFO should encourage research within organizations such as the Ecological Council of the States (ECOS), state technology assistance organizations, and other technology transfer groups to further the understanding of how to integrate environmental stewardship for nanotechnology into their ongoing assistance efforts.
- OPEI, OPPT, and ORD should support research on economic incentives for environmental stewardship behavior associated with nanomaterials and nanoproducts.
- ORD should continue to support research to promote clean production of nanomaterials and nanoproducts.

6.4. RECOMMENDATIONS FOR COLLABORATIONS

In addition to the Agency's current collaborations on nanotechnology issues and our ongoing communication activities, we recommend the following additional actions. These collaborations will reduce resource burdens on EPA's science programs and will facilitate communication with stakeholders.

- ORD should collaborate with other groups on research into the environmental applications and implications of nanotechnology. ORD's laboratories should put a special emphasis on establishing Cooperative Research and Development Agreements (CRADAs) to leverage non-federal resources to develop environmental applications of nanotechnology (CRADAs are established between the EPA and research partners to leverage personnel, equipment, services, and expertise for a specific research project.)
- EPA should collaborate with other countries (e.g., through the OECD) on research on potential human health and environmental impacts of nanotechnology.
- OCIR should lead efforts to investigate the possibilities for collaboration with and through state and local government economic development, environmental and public health officials and organizations.
- OPA and program offices, as appropriate, should lead an Agency effort to implement the communication strategy for nanotechnology.
- OPEI's Small Business Omsbudsman should engage in information exchange with small businesses, which comprise a large percentage of U.S. nanomaterial producers.

6.5. RECOMMENDATION TO CONVENE AN INTRA-AGENCY WORKGROUP

The Agency should convene a standing intra-Agency group to foster information sharing regarding risk assessment, and regulatory activities, as well as pollution prevention and stewardship-oriented activities regarding nanomaterials across program offices and regions.

6.6. RECOMMENDATION FOR TRAINING

EPA has begun educating itself about nanotechnology through seminars in the program and regional offices, an internal cross-Agency workgroup (NanoMeeters) with an extensive database, and a Millenium lecture series covering both the administrative and technical aspects of nanotechnology. The SPC Nanotechnology Workgroup also held a "primer" session on nanotechnology to help inform its members during the early stages of development of this paper. While this white paper also provides information for Agency managers and scientists, there should be ongoing education and training as well as expert support for EPA managers and staff to assist in the understanding of nanotechnology, its potential applications, regulatory and environmental implications, as well as unique considerations when conducting risk assessments on nanomaterials relative to macro-sized materials.

6.7. SUMMARY OF RECOMMENDATIONS

EPA should begin taking steps to help ensure both that society accrues the important benefits to environmental protection that nanotechnology may offer and that the Agency understands potential risks from human and environmental exposure to nanomaterials. Table 6 summarizes the staff recommendations identified above.

Table 6. Summary of Workgroup Recommendations Regarding Nanomaterials

6.1 Research for Environmental Applications. EPA should undertake, collaborate on, and support research on the various types of nanomaterials to better understand and apply information regarding their environmental applications. Specific research recommendations for each area are identified in the text.
6.2 Research for Risk Assessment. EPA should undertake, collaborate on, and support research on the various types of nanomaterials and nanotechnologies to better understand and apply information regarding: i) chemical identification and characterization, ii) environmental fate and treatment methods, iii) environmental detection and analysis, iv) potential human exposures, their measurement and control, v) human health effects assessment, vi) ecological effects assessment, and vii) conducting case studies to further identify unique risk assessment considerations for nanomaterials. Specific research recommendations for each area are identified in the text.
6.3 Pollution Prevention, Stewardship and Sustainability. EPA should engage resources and expertise as nanotechnology industries form and develop to encourage, develop and support nanomaterial pollution prevention at its source and an approach of stewardship. Detailed pollution prevention recommendations are identified in the text. Additionally, the Agency should draw on the "next generation" nanotechnologies for applications that support environmental stewardship and sustainability, such as green energy and green manufacturing.
6.4 Collaboration. EPA should continue and expand its collaborations regarding nanomaterial applications and potential human and environmental health implications.
6.5 Intra-Agency Workgroup. EPA should convene a standing intra-Agency group to foster information sharing regarding risk assessment or regulatory activities for nano-materials across program offices and regions.
6.6 Training. EPA should continue and expand its activities aimed at training Agency scientists and managers regarding potential environmental applications and environmental implications of nanotechnology.

In: Nanotechnology and the Environment   ISBN: 978-1-60692-663-5
Editor: Robert V. Neumann   © 2010 Nova Science Publishers, Inc.

Chapter 7

REFERENCES

U.S. Environmental Protection Agency

Aitken, R. J., Creely, K. S. & Tran, C. L. (2004). *Nanoparticles: An Occupational Hygiene Review*. Research Report 274. Prepared by the Institute of Occupational Medicine for the Health and Safety Executive, North Riccarton, Edinburgh, England.

Atkinson, R. (2000). Atmospheric Oxidation (Chapter 14), in Boethling, R.S.; Mackay, D. (eds.), *Handbook of Property Estimation Methods for Chemicals, Environmental and Health Sciences*, Lewis Publishers, CRC Press, Boca Raton, FL.

Ball, P. (2005). Nanomaterials Draw Electricity from Heat. *Nature Materials Update. 24* March 2005.

Ball, P. (2004). Nanotubes Show the Way to Wind Power. *Nature Materials Update. 2* September 2004.

Balshaw, D. M., Philbert, M. & Suk, W. A.. (2005). Research Strategies for Safety Evaluation of Nanomaterials, Part III: *Nanoscale Technologies for Assessing and Improving Public Health. Toxicol. Sci. 88(2):* 298-306.

Baron, P. A., Maynard, A. D. & Foley, M. (2003). Evaluation of Aerosol Release During the Handling of Unrefined Single Walled Carbon Nanotube Material. *NIOSH-DART-02-191* Rev. 1.1, April 2003.

Bekyarova, E., Ni, Y., Malarkey, E. B., Montana, V., McWilliams, J. L., Haddon, R. C. & Parpura, V. (2005). Applications of Carbon Nanotubes in Biotechnology and Biomedicine. *J. Biomedical Nanotechnology 1*:3-17.

Bidleman, T. F. (1988). Atmospheric Processes, Wet and Dry Deposition of Organic Compounds are Controlled by their Vapor-Particle Partitioning. *Environ. Sci. Technol. 22(4)*, 361-367.

Biswas P. & Wu, C. Y. (2005). Nanoparticles and the Environment. *J. Air & Waste Manage. Assoc.* 55:708-746.

Biswas, P., Yang, G. & Zachariah, M. R (1998). In Situ Processing of Ferroelectric Materials from Lead Waste Streams by Injection of Gas Phase Titanium Precursors: Laser Induced Fluorescence and X-Ray Diffraction Measurements. Combust. *Sci. Technol.134*: 183-200.

Biswas, P. & Zachariah, M. R. (1997). In Situ Immobilization of Lead Species in Combustion Environments by Injection of Gas Phase Silica Sorbent Precursors. *Environ. Sci. Technol. 31(9)*: 2455-2463.

Boethling, R. S. & Nabholz, J. V. (1997). Environmental Assessment of Polymers Under the U.S. Toxic Substances Control Act, Chapter 10. pp. 187-234. in Hamilton, J. D. and R. Sutcliffe (eds.), *Ecological Assessment of Polymers: Strategies for Product Stewardship and Regulatory Programs*. Van Nostrand Reinhold, New York. 345 p.

Borm, P., Klaessig, F. C., Landry, T. D., Moudgil, B., Pauluhn, J., Thomas, K., Trottier, R. & Wood, S. (2006). Research Strategies for Safety Evaluation of Nanomaterials, Part V: Role of Dissolution in Biological Fate and Effects of Nanoscale Particles. *Toxicol. Sciences 90(1)*: 23-32.

Borm, P. J. A. & Hreyling, W. (2004). A Need for Integrated Testing of Products in Nanotechnology, in Nanotechnologies: A Preliminary Risk Analysis on the Basis of a Workshop, Organized in Brussels on 1-2 March 2004 by the Health and Consumer Protection Directorate General of the European Commission. http://europa.eu.int/comm/health/ph_risk/events_risk_en.htm.

Boyd, A. M., Lyon, D., Velasquez, V., Sayes, D. Y., Fortner, J. & Colvin, V. L. (2005). *Photocatalytic Degradation of Organic Contaminants by Water-Soluble Nanocrystalline C60*. ACS Meeting Abstracts, 229th ACS National Meeting, San Diego, CA, March 13-17, 2005.

Brown, M. (2005a). *Nano-Bio-Info Pathways to Extreme Efficiency*. Presentation to the AAAS Annual Meeting, Washington, DC. http://www.ornl.gov/sci/eere/aaas/abstracts.htm.

Brown, M., Laitner. & J. A. (2005b). Emerging Industrial Innovations to Create New Energy-Efficient Technologies, in *Proceedings of the American Council for an Energy-Efficient Economy* (ACEE) Summer Study on Energy Efficiency in Industry, pp. 4-70 to 4-83.

Brzoska, M., Langer, K., Coester, C., Loitsch, S., Wagner, T.O. & Mallinckrodt, C. (2004). Incorporation of Biodegradable Nanoparticles into Human Airway Epithelium Cells-In vitro Study of the Suitability as a Vehicle for Drug or Gene Delivery in Pulmonary Diseases. *Biochem. Biophys. Res. Commun. 318(2)*: 562-570.

Cai R. & et al. (1992). Increment of Photocatalytic Killing of Cancer Cells Using TiO_2 with the Aid of Superoxide Dismutase. *The Chemical Society of Japan, Chemistry Letters*: 427-430.

CBEN. (2005). Center for Biological and Environmental Nanotechnology, Rice University. Information about the center and current research summaries are available online: http://cohesion.rice.edu/centersandinst/cben/.

Chen, B. & Beckett, R. 2001). Development of SdFFF-ETASS for Characterizing Soil and Sed. *Colloids Analyst 126*:1588-1593.

Chen, B. & Selegue, J. (2002). Separation and Characterization of Single-Walled and Multiwalled Carbon Nanotubes by Using Flow Field-Flow Fractionation. *Anal. Chem. 74* (18): 4774-4780.

Chen, C., Sheng, G., Wang, X., Fu, J., Chen, J. & Liu, S. (2000). Adsorption Characteristics of Fullerenes and Their Application for Collecting VOCs in Ambient Air. Juanjing Juaxue, 19(2), 165-169. [The original report is published in Chinese. The abstract published in Chemical Abstracts does not specify if the fullerenes used are free particles or immobilized.]

Chen, Y., Crittenden, J. C., Hackney, S., Sutter, L. & Hand, D. W. (2005). Preparation of a Novel TiO_2-Based p-n Junction Nanotube Photocatalyst. *Environ. Sci. Technol. 39(5)*: 1201-1208

Cheng, S.H. & Cheng, J. (2005). *Carbon Nanotubes Delay Slightly the Hatching Time of Zebrafish Embryos*. 229th American Chemical Society Meeting, San Diego, CA March 2005.

Cheng, X., Kan, A. T. & Tomson, M. B. (2004). Naphthalene Adsorption and Desorption from Aqueous C60 Fullerene. *J. Chem. Eng. Data 49*: 675-683.

Christen, K. (2004). Novel Nanomaterial Strips Contaminants from Waste Streams. *Environ. Sci. Technol. 38(23)*: 453A-454A.

Colvin, V. (2003). The Potential Environmental Impact of Engineered Nanoparticles. *Nature Biotechnol. 21(10)*, 1166-1170.

Comparelli, R., Cozzoli, P. D., Curri, M. L., Agostiano, A., Mascolo, G. & Lovecchio, G. (2004). Photocatalytic Degradation of Methyl-red by Immobilized Nanoparticles of TiO2 and ZnO. *Water Sci. Technol. 49(4)*: 183-188.

Das R., Kiley, P. J., Segal, M., Norville, J., Yu, A. A., Wang, L., Trammell, S. A., Reddick, L. E., Kumar, R., Stellacci, F., Lebedev, N., Schnur, J., Bruce, B. D, Zhang, S. & Baldo, M. (2004). Integration of Photosynthetic Protein Molecular Complexes in Solid-State Electronic Devices. *Nano Letters: 4(6)*: 1079-1083.

Dennekamp, M., Mehenni, O. H., Cherrie, J. & Seaton, A. (2002). Exposure to Ultrafine Particles and PM2.5 in Different Micro-Environments. *Annals of Occupational Hygiene 46 (suppl. 1)*: 412–414.

Derfus, A. M., Chan, W. C. W. & Bhatia, S. N. (2004). Probing the Cytotoxicity of Semiconductor Quantum Dots. *Nano Letters 4(1)*:11-18.

Diallo, M. S., Christie, S., Swaminathan, P., Johnson, J. H., Jr. & Goddard, W. A., III. (2005). Dendrimer Enhanced Ultrafiltration. 1. Recovery of Cu (II) from Aqueous Solutions Using PAMAM Dendrimers with Ethylene Diamine Core and Terminal NH2 Groups. *Environ. Sci. Technol. 39(5)*: 1366-1377

Dick, K. A., Deppert, K., Larsson, M. W., Mårtensson, T. Seifert,W., Wallenberg, L. R. & Samuelson, L. (2004). Synthesis of branched 'nanotrees' by controlled seeding of multiple branching events. *Nature Materials 3*: 380–384.

Donaldson, K., Aitken, R., Tran, L., Stone, R., Duffin, R., Forrest, G. & Alexander, A. (2006). Carbon Nanotubes: A review of Their Properties in Relation to Pulmonary Toxicology and Workplace Safety. *Toxicol. Science 92*:5-22.

Dreher, K. L (2004). Health and Environmental Impact of Nanotechnology: Toxicological Assessment of Manufactured Nanoparticles. *Toxicological Sciences 77*:3-5.

Dror, I., Baram, D. & Berkowitz, B. (2005). Use of Nanosized Catalysts for Transformation of Chloro-Organic Pollutants. Environ. *Sci. Technol. 39(5)*: 1283-1290.

Elliott & et al. (2005). Novel Products From the Degradation of Lindane by Nanoscale Zero Valent Iron. American Chemical Society Annual Meeting, San Diego, CA, Abstract.

European Commission Scientific Committee on Emerging and Newly Identified Health Risks (SCENIHR). (2006). The Appropriateness of Exisiting Methodolgies to Assess the Potential Risks Associated with Engineered and Adventitious Products of Nanotechnologies. Document number SCENIHR/002/05.

European Commission. (2004). European Commission, Community Health and Consumer Protection. Nanotechnologies: A Preliminary Risk Analysis on the Basis of a Workshop

Organized in Brussels on 1-2 March 2004 by the Health and Consumer Protection Directorate General of the European Commission. http://europa.eu.int/comm/health/ ph_risk/events_risk_en.htm.

European NanoSafe Report. (2004). *Technical Analysis: Industrial Application of Nanomaterials Chances and Risks.* www.nano.uts.edu.au/nanohouse/ nanomaterials%20risks.pdf.

Filley, T. R., Ahn, M., Held, B. W. & Blanchette, R. A. (2005). Investigations of Fungal Mediated (C60-C70) Fullerene Decomposition. Preprints of Extended Abstracts Presented at the ACS National Meeting, *American Chemical Society, Division of Environmental Chemistry 45(1)*, 446-450.

Fortner, J. D., Lyon, D. Y., Sayes, C. M., Boyd, A. M, Falkner, J. C., Hotze, E. M., Alemany, L. B, Tao, Y. J., Guo, W., Ausman, K. D., Colvin, V. L. & J. B. Hughes. (2005). C60 in water: Nanocrystal Formation and Microbial Response. *Environ. Sci. Technol. 39*:4307-4316.

Frampton, M. W., Utell, M. J., Zareba, W., Oberdörster, G., Cox, C., Huang, L. S., Morrow, P. E., Lee, F. E. H., Chalupa, D., Frasier, L. M., Speers, D. M. & Stewart. J. (2004). Effects of Controlled Exposure to Ultrafine Carbon Particles in Healthy Subjects and Subjects with Asthma. *Health Effects Institute. Report 126*: 1-47.

Frazer, L. (2003). Organic Electronics: A Cleaner Substitute for Silicon. Environ. *Health Perspect.* 111:5.

Frink, C. R., Waggoner, P. E. & Asubel, J. H. (1996). Nitrogen Fertilizer: Retrospect and Prospect. *Proc. Natl. Acad. Sci.*, p. 1175-1180.

Gardner, P., Hofacre, K. & Richardson, W. (2004). Comparison of Simulated Respirator Fit Factors Using Aerosol and Vapor Challenges. *J. Occupat. Environ. Hygiene 1*: 29–38.

Georgia Tech. (2005). March 2005 press release http://gtresearchnews.gatech.edu/newsrelease/adhesive.htm; Abstract posted at: http://cfpub.epa.gov/ncer_abstracts/index.cfm/fuseaction/display.abstractDetail/abstract/6 352/rep ort/0

Grassian, V. H., O'Shaughness, P. T, Adamcakova-Dodd, A., Pettibone, J. M. & Thorne, P. S. (2006). unpublished results

Gurevich, L., Canali, L. & Kouwenhoven, L. P. (2000). Scanning gate spectroscopy on nanoclusters. *Applied Physics Letters 76(3)*:384.

Hardman, R. (2006). A Toxicological Review of Quantum Dots: Toxicity Depends on Physicochemical and Environmental Factors. Environ. *Health Perspect. 114(2):* 165-172.

Health Effects Institute, Communication 9, August (2001). www.healtheffects.org/pubs¬ comm.htm.

Hinds, W.C. (1999). *Aerosol Technology: Properties, Behavior, and Measurement of Airborne Particles.* 2nd ed. John Wiley and Sons, Inc., New York.

Höhr, D., Steinfartz, Y., Schins, R. P. F., Knaapen, A. M., Martra, G., Fubini , B. &. Borm, P. J. A. (2002). The Surface Area Rather Than the Surface Coating Determines the Acute Inflammatory Response After Instillation of Fine and Ultrafine TiO2 in the Rat. *Int. J. Hyg. Environ. Health 205*:239-244.

Harder, V., Gilmour, P., Lentner, B., Karg, E., Takenaka, S., Ziesenis, A., Stampfl, A., Kodavanti, U., Heyder, J. & Schulz, H. (2005). Cardiovascular Responses in Unrestrained WKY Rats to Inhaled Ultrafine Carbon Particles. *Inhal. Toxicol. 17*:29-42.

Holsapple, M. P., Farland, W. H., Landry, T. D., Monteiro-Riviere, N. A., Carter, J. M., Walker N. J. & Thomas, K. V. (2005). Research Strategies for Safety Evaluation of Nanomaterials, Part II: Toxicological and Safety Evaluation of Nanomaterials, Current Challenges and Data Needs. *Toxicol. Sci. 88(1)*: 12-17.

Hu, J., Lo, I. M. & Chen, G. (2004). Removal of Cr(VI) by Magnetite Nanoparticle. Water Sci. *Technol. 50(12)*:139-46.

Hughes L. S., Cass, G. R., Gone, J., Ames, M. & Olmez, I. (1998). Physical and Chemical Characterization of Atmospheric Ultrafine Particles in the Los Angeles Area. *Environ. Sci. Technol. 32(9)*:1153-1161.

Ivanov, V., Tay, J. H., Tay, S. T. & Jiang, H. L. (2004). Removal of Micro-Particles by Microbial Granules used for Aerobic Wastewater Treatment. *Water Sci. Technol. 50(12)*: 147-154.

Jia, G., Wang, H.,Yan, L., Wang, X., Pei, R., Yan, T., Zhao, Y. & Guo, X. (2005). Cytotoxicity of Carbon Nanomaterials. *Environ. Sci. Technol. 39*:1378-1383.

Kanel, S. R., Manning, B., Charlet, L. & Choi, H. (2005). Removal of Arsenic (III) from Groundwater by Nanoscale Zero-Valent Iron. *Environ. Sci. Technol. 39(5)*:1291-1298.

Kreyling, W. G., Semmler, M., Erbe, F., Mayer, P., Takenaka, S., Schulz, H., Oberdörster., G. & Ziesenis, A. (2002). Translocation of Ultrafine Insoluble Iridium Particles From Lung Epithelium to Extrapulmonary Organs is Size Dependent But Very Low. *J. Toxicol. Environ. Health A 65*:1513-1530.

Kulkarni P., Namiki N., Otani Y. & Biswas P. (2002) Charging of particles in unipolar coronas irradiated by in-situ soft X-rays: Enhancement of Capture Efficiency of Ultrafine Particles. *J. Aerosol Sci. 33 (9)*, 1279-1298

Lam, C. W., James, J. T., McCluskey, R. & Hunter, R. L. (2004). Pulmonary Toxicity of Single-Walled Carbon Nanotubes in Mice 7 and 90 Days after Intratracheal Instillation. *Toxicol. Sci. 77*:126-134.

Lecoanet, H. F., Bottero, J. Y. & Wiesner, M. R. (2004). Laboratory Assessment of the Mobility of Nanomaterials in Porous Media. *Environ. Sci. Technol. 38*:5164-5169.

Lecoanet, H. F. & Wiesner, M. R. (2004). Velocity Effects on Fullerene and Oxide Nanoparticle Deposition in Porous Media. *Environ. Sci. Technol. 38*:4377-4382.

Lee, M. H., Cho, K., Shah, A. P. & Biswas, P. (2005). Nanostructured Sorbents for Capture of Cadmium Species in Combustion Environments. *Environ. Sci. Technol. 39(21)*:8481-8489.

Li, X. Y., Brown, D., Smith S., MacNee, W. & Donaldson, K. (1999). Short Term Inflammatory Responses Following Intratracheal Instillation of Fine and Ultrafine Carbon Black in Rats. *Inhal. Toxicol. 11*:709-731.

Lloyd, S. M., Lave, L. B. & Matthews, H. S. (2005). Life Cycle Benefits of Using Nanotechnology to Stabilize Platinum-Group Metal Particles in Automotive Catalysts. *Environ. Sci. Technol. 39*:1384-1392.

Lockman P. R., Kozaria, J. M., Mumper, R. J. & Allen, D. D. (2004). Nanoparticle Surface Charges Alter Blood-Brain Barrier Integrity and Permeability. *J. Drug Targeting 12 (9-10)*:635-641.

Lovern, S. B. & Klaper, R. (2006) *Daphnia magna* mortality when exposed to titanium dioxide and fullerene (C60) nanoparticles. *Environ. Toxicol. Chem. 25(4)*:1132-1137.

Luther, W., ed. (2004). *Technological Analysis, Industrial Application of Nanomaterials - Chances and Risks.* Future Technologies Division, VDI Technologiezentrum GmbH, Düsseldorf, Germany.

Lux Research. (2006). http://www.luxresearchinc.com/TNR4_TOC.pdf http://www. luxresearchinc.com/press/RELEASE_TNR4.pdf.

Lux Research. (2004). www.luxresearchinc.com/press/RELEASE_econ.pdf

Madan, T., Munshi, N., De, T. K., Usha Sarma, P. & Aggarwal, S. S. (1997). Biodegradable Nanoparticles as a Sustained Release System for the Antigens/Allergens of Aspergillus fumigatus: Preparation and Characterization. *Int. J. Pharm. 159*, 135-147.

Malik, N., Wiwattanapatapee, R., Klopsch, R., Lorenz, K., Frey, H., Weener, J. W., Meijer, E. W., Paulus, W. & Duncan, R. (2000). Dendrimers: Relationship Between Structure and Biocompatibility In Vitro, and Preliminary Studies on the Biodistribution of 125I-Labeled Polyamidoamine Dendrimers In Vivo. *J. Control. Release 65*:133-148.

Wang, W. X. & Masciangioli, T., (2003). Environmental Technologies at the Nanoscale. *Environ. Sci. Technol. A-Pages. 37(5)*:102A-108A.

Maynard, A. D. & Kuempel, E. D. (2005). Airborne Nanostructured Particles and Occupational Health. *J. Nanoparticle Res. 7(6)*: 587-624.

Maynard, A. D., Baron, P. A., Foley, M., Shvedova, A. A., Kisin, E. R. & Castranova, V. (2004). Exposure to Carbon Nanotube Material: Aerosol Release During the Handling of Unrefined Single-Walled Carbon Nanotube Material. *J. Toxicol. Environ. Health A 67*: 87-107.

Maynard, A. D. (2000). Overview of Methods for Analysing Single Ultafine Particles. *Phil. Trans. R. Soc. Lond. A 358*: 2593-2610.

McKim, J., Schmieder, P. & Veith, G. (1985). Adsorption Dynamics of Organic Chemical Transport Across Trout Gills as Related to Octanol-Water Partition Coefficient. Government Reports Announcements and Index, Issue 17, NTIS report number PB85-198315, 12 p.

McMurry, P. H. (2000). A Review of Atmospheric Aerosol Measurements. *Atmospheric Environ. 34(12-14)*:1959-1999.

Monteiro-Riviere, N. A., Nemanich, R. J., Inman, A. O., Yunyu, Y. W. & Riviere, J. E. (2005). Multi-walled carbon nanotubes interactions with human epidermal keratinocytes. *Toxicol. Lett. 155(3)*: 377-384.

Moore, M. N. (2006). Do Nanoparticles Present Ecotoxicological Risks for the Health of the Aquatic Environment? *Environ. Int. 32(8)*: 967-976.

Morgan, K. (2005). Development of a Preliminary Framework for Informing the Risk Analysis and Risk Management of Nanoparticles. *Risk Anal. 25*:1-15.

Muller, J., Huaux, F., Moreau, N., Misson, P., Heilier, J. F., Delos, J., Arras, M., Fonseca, A., Nagy, J. B. & Lison, D. (2005). Respiratory toxicity of multi-wall carbon nanotubes. *Toxicol. Appl. Pharmacol. 207*: 221-231.

Murashov. V. (2006). Letter to the Editor:Comments on *Particle Surface Characteristics May Play an Important Role in Phytotoxicity of Alumina Nanoparticles by Yang, L., Watts, D.J., Toxicology Letters, 2005, 158, 122-132.* Toxicol. Lett 164 185-187.

National Institute for Occupational Health and Safety. (2005)a. Approaches to Safe nanotechnology: An Information Exchange with NIOSH. http:www.cdc.gov/ niosh/topics/nanotech/nano_exchnge.html.

National Institute for Occupational Health and Safety. (2005b). Strategic Plan for NIOSH Nanotechnology Research Program. http:www.cdc.gov/niosh/topics/nanotech/strat_ plan.html.

National Institute for Occupational Health and Safety. (2004). Nanotechnology Workplace Safety and Health. http://www.cdc.gov/niosh/topics/nanotech/default.html.

National Institute for Occupational Health and Safety. (2003). Filtration and Air-Cleaning Systems to Protect Building Environments. Cincinnati, OH: U.S. Department of Health and Human Services, Public Health Service, Centers for Disease Control and Prevention, National Institute for Occupational Safety and Health, DHHS (NIOSH) Publication No. 2003–136.

National Research Council. (1983). *Risk Assessment in the Federal Government: Managing the Process.* National Academy of Sciences, Washington, D.C. 192 pp.

National Research Council. (1994). Science and Judgment in Risk Assessment, National Academy of Sciences, Washington, D.C.

National Nanotechnology Initiative. (2006a). What is Nanotechnology? http://www.nano.gov/html/facts/home_facts.html

National Nanotechnology Initiative. (2006b). About the NNI. http://www.nano.gov/html/about/home_about.html

National Nanotechnology Initiative. (2006c). Environmental, Health and Safety Research Needs for Engineered Nanoscale Materials. http://www.nano.gov.

National Nanotechnology Initiative. (2004). National Nanotechnology Initiative Strategic Plan, Goal 4: Support Responsible Development of Nanotechnology. http://www.nano.gov/NNI_Strategic_Plan_(2004).pdf.

National Nanotechnology Initiative. (2000). The Initiative and Its Implementation Plan. http://www.nano.gov/html/facts/whatIsNano.html.

Nagaveni, K., Sivalingam, G., Hegde, M. S. & Madras, G. (2004). Photocatalytic Degradation of Organic Compounds over Combustion-Synthesized Nano-TiO2. *Environ. Sci. Technol. 38*, 1600¬1604.

Nel, A., Xia, T., Madler, L. & Li, N. (2006). Toxic Potential of Materials at the Nanolevel. *Science 311*: 622-627.

Nemmar A., Hoylaerts, M. F., Hoet, P. H. M., Vermylen, J. & Nemery, B. (2003). Size Effect of of Intratracheally Instilled Particles on Pulmonary Unflammation and Thrombosis. *Toxicol. Appl. Pharmacol. 186*: 38-45.

Nigavekar, S. S., Sung, L. Y., Llanes, M., El-Jawahri, A., Lawrence, T. S., Becker, C. W., Blaogh, L. & Khan, M. K. (2004). 3H Dendrimer Nanoparticle Organ/Tumor Distribution. *Pharm. Res. 21 (3)*:476-483.

Niimi, A. & Oliver, B. (1988). Influence of Molecular Weight and Molecular Volume on Dietary Adsorption Efficiency of Chemicals by Fishes. *Can. J. Fish. Aquat. Sci. 45(2)*:222-227.

NREL. (2005). National Renewable Energy Laboratory Nanoscience & Nanotechnology: Meeting 21st Century Energy Challenges.

Nurmi, J. T., Tratnyek, P. G., Sarathy, V., Baer, D. R., Amonette, J. E., Pecher, K., Wang, C., Linehan, J. C., Matson, D. W., Penn, R. L. & Driessen, M. D. (2005). Characterization and Properties of Metallic Iron Nanoparticles: Spectroscopy, Electrochemistry, and Kinetics. *Environ. Sci. Technol. 39(5)*:1221-1230.

Oberdörster, E. (2004b). Manufactured nanomaterial (fullerenes, C60) induce oxidative stress in the brain of juvenile largemouth bass. *Environ. Health Perspect. 12(10):*1058-1062

Oberdörster E. (2004c). Toxicity of nC$_{60}$ Fullerenes to Two Aquatic Species: Daphnia and Largemouth bass. American Chemical Society, Anaheim, CA, March 27-April 2004. Abstract IEC21

Oberdörster, G., Oberdörster, E., Oberdörster, J. (2005a). Nanotoxicology: An Emerging Discipline Evolving from Studies of Ultrafine Particles. *Environ. Health Perspect. 113(7)*: 823-839.

Oberdörster, G., Maynard, A., Donaldson, K., Castranova, V., Fitzpatrick, J., Ausman, K., Carter, J., Karn, B., Kreyling, W., Lai, D., Olin, S., Monteiro-Riviere, N., Warheit, D. & Yang, (2005)b. Principles for characterizing the potential human health effects from exposure to nanomaterials: elements of a screening strategy. A report from the ILSI Research Foundation/Risk Science Institute Nanomaterial Toxicity Screening Working Group. Part. *Fibre Toxicol.: 2*:8.

Oberdörster, G., Sharp, Z., Atudorei, V., Elder, A., Gelein, R., Kreyling, W. & Cox, C. (2004a). Translocation of inhaled ultrafine particles to the brain. *Inhal. Toxicol. 16*:437-445.

Oberdörster, G., Sharp, Z., Atudorei, V., Elder, A., Gelein, R., Lunts, A., Kreyling, W. & Cox, C., (2002). Extrapulmonary translocation of ultrafine carbon particles following whole-body inhalation exposure of rats. *J. Toxicol. Environ. Health A 65*:1531-1543.

Oberdörster, G. (1996). Significance of Particle Parameters in the Evaluation of Exposure-Dose-Response Relationships of Inhaled Particles. *Inhal. Toxicol. 8* (Suppl. 8):73-89.

Oberdörster G., Ferin, J. & Lehnert, B. E. (1994). Correlation Between Particle Size, In Vivo Particle Persistence, and Lung Injury. *Environ. Health Perspect. 102*(Suppl 5):173-179.

Organisation for Economic Co-operation and Development. (2001). Environmental Strategy for the First Decade of the 21st Century. Adopted by OECD Environment Ministers. 16 May 2001. http://www.oecd.org/dataoecd/33/40/1863539.pdf.

Opperhuizen, A., Velde, E., Gobas, F., Llem, D. & Steen, J. (1985). Relationship between bioconcentration in fish and steric factors of hydrophobic chemicals. *Chemosphere 14(11/12)*:1871-1896.

Oxonica, (2005). Envirox Fuel Efficiency Promotional Literature. http://www.oxonica.com/cms/promotional/Fuel-Efficiency.pdf.

Pickering, K. D., Wiesner & M. R. (2005). Fullerol-Sensitized Production of Reactive Oxygen Species in Aqueous Solution. *Environ. Sci. Technol. 39(5)*:1359-1365.

Pitoniak, E., Wu, C. Y., Mazyck, D. W., Powers, K. W. & Sigmund, W. (2005). Adsorption enhancement mechanisms of silica-titania nanocomposites for elemental mercury vapor removal. *Environ. Sci. Technol. 39(5)*:1269-1274.

Powers, K. W., Brown, S. C, Krishna, V. B., Wasdo, S. C., Moudgil, B. M. & Robert, S. M (2006). Research Strategies for Safety Evaluation of Nanomaterials, Part VI: Characterization of Nanoscale Particles for Toxicological Evaluation. *Toxicol. Sci. 90(2)*: 296-303.

Preining, O. (1998). The Physical Nature of Very, Very Small Particles and its Impact on Their Behaviour. *J. Aerosol Sci. 29(5/6)*:481-495.

Quinn, J., Geiger, C., Clausen, C., Brooks, K., Coon, C., O'Hara, S., Krug, T., Major, D., Yoon, W. S., Gavsakar, A. & Holdsworth, T. (2005). Field Demonstration of DNAPL

Dehalogenation Using Emulsified Zero-Valent Iron. *Environ. Sci. Technol. 39(5)*:1309-1318.

Reguera, G., McCarthy, K. D., Mehta, T., Nicoll, J. S., Tuominen, M. T. & Lovley, D. R. (2005). Extracellular Electron Transfer Via Microbial Nanowires. *Nature 453(23)*:1098-1101.

Renwick, L. C., Donaldson, K. & Clouter, A. (2001). Impairment of Alveolar Macrophage Phagocytosis by Ultrafine Particles. Toxicol. *Appl. Pharmacol. 172*:119-127.

Roberts, D.W. & et al. (2005). Localization of Intradermally Injected Quantum Dot Nanoparticles in Regional Lymph Nodes. Society of Toxicology Annual Meeting, New Orleans, LO, 2005, Abstract.

Roberts, S. M. (2005). Developing Experimental Approaches for the Evaluation of Toxicological Interactions of Nanoscale Materials. Workshop proceedings, University of Florida, Gainesville, FL. Nov. 3-4, 2004. http://ntp.niehs.nih.gov/files/NanoTox Workshop.pdf, http://www.nanotoxicology.ufl.edu/workshop/index.html.

Rocha J. C., de Sen, J. J., dos Santos, A., Tosacano, I. A. S. & Zara, L. F. (2000). Aquatic Humus from an Unpolluted Brazillian Dark-Brown Stream: General Characterization and Size Fractionation of Bound Heavy Metals. *J. Env. Monit. 2*:39-44.

Rupprecht & Patashnick Co., Inc. (2005). TEOM Series 7000 Source Particulate Monitor. Web site May 2005. http://www.rpco.com/products/cemprod/cem7000/index.htm.

Ryman-Rasmussen, J. P., Riviere, J. E. & Monteiro-Riviere, N. A. (2006). Penetration of Intact Skin by Quantum Dots with Diverse Physicochemical Properties. *Toxicol. Sci. 91(1)*:159-165.

Sayes, C. M., Wahi, R., Kurian, P. A., Liu, Y., West, J. L., Ausman, K. D., Warheit, D. B., & Colvin, V. L. (2006). Correlating Nanoscale Titania Structure with Toxicity: A Cytotoxicity Inflammatory Reponse Study with Human Dermal Fibroblasts and Human Lung Epithelial Cells. *Toxicol. Sci. 92(1)*:174-185.

Sayes, C. M., Fortner, J. D., Guo, W., Lyon, D., Boyd, A. M., Ausman, K. D., Tao, Y. J., Sitharaman, B., Wilson, L. J., Hughes, J. B., West, J. L. & Colvin, V. L. (2004). The Differential Cytotoxicity of Water Soluble Fullerenes. *Nano Letters 4(10)*:1881-1887, 2004

Schwarzenbach, R. P., Gshwend, P. M. & Imboden, D. M., (eds.) (1993). Sorption: Solid-Aqueous Solution Exchange (Chapter 11) in *Environmental Organic Chemistry*, Wiley-Interscience, New York.

Sclafani, A. & Herrmann, J. M. (1996). Comparison of the Photoelectronic and Photocatalytic Activities of Various Anatase and Rutile Forms of Titania in Pure Liquid Organic and in Aqueous Phases. *J. Phys. Chem. 100*:13655-13661.

Shvedova, A. A., Kisin, E. R., Mercer, R., Murray, A. R., Johnson, V. J., Potapovich, A. I., Tyurina, Y. Y., Gorelik, O., Arepalli, S., Schwegler-Berry, D., Hubbs, A. F., Antonini, J., Evans, D. E., Ku, B. K., Ramsey, D., Maynard, A., Kagan, V. E., Castranova, V. & Baron, P. (2005). Unusual Inflammatory and Fibrogenic Pulmonary Responses to Single Walled Carbon Nanotubes in Mice. *Am. J. Physiol. Lung Cell Mol. Physiol. 289*:L698-L708.

Shvedova, A. A., Castranova, V., Kisin, E. R., Scwegler, B. D., Murray, A. R., Gandelsman, V. Z., Maynard, A. & Baron, P. (2003). Exposure to Carbon Nanotube Material: Assessment of Nanotube Cytotoxicity using Human Keratinocyte Cells. *J. Toxicol.*

Environ. Health A. 66(20): 1909-1926. Small Times Media, LLC, Nanotechnology Products Report, August 2005.

Spurny, K. R. (1998). On the Physics, Chemistry and Toxiology of Ultrafine Anthropogenic, Atmospheric Aerosols (UAAA): New Advances. *Toxicol. Lett. 96-97*: 253-261.

Stevens, G. & Moyer, E. (1989). 'Worst case' aerosol testing parameters: I. Sodium chloride and dioctyl phthalate aerosol filter efficiency as a function of particle size and flow rate. Am. Indust. *Hygiene Assoc. J. 50(5)*:257-64.

Steinfeldt, M., Petschow, U., Haum, R. & Gleich, A. V. (2004). Nanotechnology and Sustainability. Discussion Paper #65/04. Institute for Ecological Economy Research. Berlin. www.ioew.de.

Sun, J. D., Wolff, R. K. & Kanapilly, G. M. (1982). Deposition, Retention and Biological Fate of Inhaled Benzo(a)pyrene Adsorbed onto Ultrafine Particles and as a Pure Aerosol. Toxicol. *Appl. Pharmacol. 65(2)*: 231-244.

Sun, J. D., Wolff, R. K., Kanapilly, G. M. & McClellan, R. O. (1984). Lung Retention and Metabolic Fate of Inhaled Benzo(a)pyrene Associated with Diesel Exhaust Particles. Toxicol. *Appl. Pharmacol. 73(1)*: 48-59.

Swiss Report Reinsurance Company. (2004). Nanotechnology: Small Matter, Many Unknowns. www.swissre.com.

Thomas, K. & Sayre, P. (2005). Research Strategies for Safety Evaluation of Nanomaterials, Part I: Evaluating Human Health Implications for Exposure to Nanomaterials. *Toxicol.Sci. 87(2)*: 316-321.

Tinkle, S. S, Antonini, J. M., Rich, B. A., Roberts, J. R., Salmen, R., DePree, K. & Adkijns, E.J. (2003). Skin as a Route of Exposure and Sensitization in Chronic Beryllium Disease. *Environ. Health Perspect. 111*:1202-1208.

Tsuji, J. S., Maynard, A. D., Howard, P. C., James, J. T., Lam, C. W., Warheit, D. B. & Santamaria, A. B. (2006). Research Strategies for Safety Evaluation of Nanomaterials, Part IV: Risk Assessment of Nanoparticles. *Toxicol. Sci. 88(1)*:12-17.

Tungittiplakorn, W., Cohen, C. & Lion, L. W. (2005). Engineered Polymeric Nanoparticles for the Bioremediation of Hydrophobic Contaminants. *Environ. Sci. Technol. 39*:1354-1358.

Tungittiplakorn, W., Lion, L. W., Cohen, C. & Kim, J. Y. (2004). Engineered Polymeric Nanoparticles for Soil Remediation. *Environ. Sci. Technol. 38*: 1605-1610.

Uchino, T., Tokunaga, H., Ando, M. & Utsuni, H. (2002). Quantitative Determination of OH Radical Generation and its Cytotoxicity Induced by TiO2-UVA Treatment. *Toxicol. In Vitro 16*:629-635.

UK Department for Environment, Food and Rural Affairs. (2005) Characterising the Potential Risks Posed by Engineered Nanoparticles: A First UK Government Research Report. Available at: www.defra.gov.uk/environment/nanotech/nrcg/pdf/nanoparticles-riskreport.pdf.

UK Health & Safety Executive. (2004). Nanoparticles: An Occupational Hygiene Review. Research Report 274. http://www.hse.gov.uk/research/rrhtm/rr274.htm.

UK Royal Society. (2004. The Royal Society and the Royal Academy of Engineering. Nanoscience and Nanotechnologies: Opportunities and Uncertainties. http://www.nanotec.org.uk/finalreport.htm.

U.S. Department of Agriculture. (2003). Nanoscale Science and Engineering for Agriculture and Food Systems. Report Submitted to Cooperative State Research, Education, and

Extension Service. Norman Scott (Cornell University) and Hongda Chen (CSREES/USDA) Co-chairs.

U.S. Environmental Protection Agency. Innovation Action Council. 2005. Presentation by Jay Benforado. June 30, (2005).

U.S. Environmental Protection Agency. (2005). Office of Pollution Prevention and Toxics. 12 Principles of Green Chemistry. http://www.epa.gov/greenchemistry/principles.html.

U.S. Environmental Protection Agency. (2004). Office of Research and Development. Air Quality Criteria for Particulate Matter. Report Number EPA/600/P-99/002a,bF. October. http://cfpub2.epa.gov/ncea/cfm/recordisplay.cfm?deid=87903.

U.S. Environmental Protection Agency. (2003). Office of Water. Methodology for Deriving Ambient Water Quality Criteria for the Protection of Human Health (2000) Technical Support Document Volume 2: Development of National Bioaccumulation Factors.

U.S. Environmental Protection Agency. (1986). Health Effects Assessment for Asbestos. Washington, D.C. EPA/540/1-86/049. NTIS PB86134608.

U.S. Environmental Protection Agency. (1998). Guidelines for Ecological Risk Assessment. EPA/630/R095/002F. http://cfpub.epa.gov/ncea/raf/recordisplay.cfm?deid=12460.

U.S. Environmental Protection Agency. (1996). Health Effects of Inhaled Crystalline and Amorphous Silica. EPA/600/R-95/115.

Warheit, D. B., Webb, T. R., Sayes, C. M., Colvin, V. L. & Reed K. L. (2006). Pulmonary Instillation Studies with Nanoscale TiO_2 Rods and Dots: Toxicity Is Not Dependent Upon Particle Size and Surface Area. *Toxicol. Sci. 91(1)*: 227-236.

Warheit, D. B. , Brock, W. J., Lee, K. P., Webb, T. R. & Reed, K. L. (2005). Comparative Pulmonary Toxicity Instillation and Inhalation Studies with Different TiO2 particle Formulaitons: Impact of Surface Treatment on Particle Toxicity. *Toxicol. Sci. 88(2)*: 514-524.

Warheit, D. B, Laurence, B. R., Reed, K. L., Roach, D. H., Reynolds, G. A. & Webb, T. R. (2004). Comparative Pulmonary Toxicity Assessment of Single-wall Carbon Nanotubes in Rats. *Toxicol. Sciences 77*:117-125.

Wiesner, M. R., Lowry, G. V., Alvarez, P., Dionysiou, D. & Bisawas, P. (2006). Assessing the Risks of Manufactured Nanomaterials. *Environ. Sci. Tech. 40(14)*:4336-4345.

Willis, R.S. (2002). When Size Matters. Today's Chemist at Work, *American Chemical Society, July 2002*, p. 21-24.

Woodrow Wilson Center Project on Emerging Nanotechnologies, Inventory of Consumer Products. (2006). http://www.nanotechproject.org/44.

World Resources Institute. (2000). The Weight of Nations: Material Outflows from Industrial Economies.

Yang, L., Watts & D. J. (2005). Particle surface characteristics may play an important role in phytotoxicity of alumina nanoparticles. *Toxicol. Lett. 158*:122-132.

Zhang, T. W., Boyd,S.,Vijayaraghavan, A. & Dornfeld, D. (2006). Energy Use in Nanoscale Manufacturing. Proceedings of the 2006 IEEE International Symposium on Electronics and the Environment, pp. 266-271.

Zhang, W. (2003). Nanoscale Iron Particles for Environmental Remediation: An Overview. *J. Nanoparticle Res. 5*: 323-332.

Zhao, X., Striolo, A. & Cummings, P. T. (2005). C60 Binds to and Deforms Nulceotides. *Biophysical J. 89*:3856-3862.

Zitko V. (1981). Uptake and excretion of chemicals by aquatic fauna. pages 67 to 78 in Stokes PM (ed.) *Ecotoxicology and the Aquatic Environment*, Pergamon Press.

Conversation with Hongda Chen. May, 2005.

In: Nanotechnology and the Environment
Editor: Robert V. Neumann

ISBN: 978-1-60692-663-5
© 2010 Nova Science Publishers, Inc.

Appendix A

GLOSSARY OF NANOTECHNOLOGY TERMS

U.S. Environmental Protection Agency

Aerosol: A cloud of solid or liquid particles in a gas.

Array: An arrangement of sensing elements in repeating or non-repeating units that are arranged for increased sensitivity or selectivity.

Biomimetic: Imitating nature and applying those techniques to technology.

Buckyball/C$_{60}$: see Fullerenes, of which "buckyballs" is a subset. The term "buckyball" refers only to the spherical fullerenes and is derived from the word "Buckminsterfullerene," which is the geodesic dome / soccer ball-shaped C$_{60}$ molecule. C$_{60}$ was the first buckyball to be discovered and remains the most common and easy to produce.

Catalyst: A substance, usually used in small amounts relative to the reactants, that modifies and increases the rate of a reaction without being consumed or changed in the process.

Dendrimers: artificially engineered or manufactured molecules built up from branched units called monomers. Technically, a dendrimer is a branched polymer, which is a large molecule comprised of many smaller ones linked together.

Diamondoid: Nanometer-sizes structures derived from the diamond crystal structure.

Electron beam lithography: Lithographic patterning using an electron beam, usually to induce a change in solubility in polymer films. The resulting patterns can be subsequently transferred to other metallic, semiconductor, or insulating films.

Engineered/manufactured nanomaterials: Nanosized materials are purposefully made. These are in contrast to incidental and naturally occurring nanosized materials. Engineering/manufacturing may be done through certain chemical and / or physical processes to create materials with specific properties. There are both "bottom-up" processes (such as self-assembly) that create nanoscale materials from atoms and molecules, as well as "top-down" processes (such as milling) that create nanoscale materials from their macro-scale counterparts. Nanoscale materials that have macro-scale counterparts frequently display different or enhanced properties compared to the macro-scale form.

Exposure assessment: The determination or estimation (qualitative or quantitative) of the magnitude, frequency, duration, route, and extent (number of people) of exposure to a chemical, material, or microorganism.

Fullerenes: Pure carbon, cage-like molecules composed of at least 20 atoms of carbon. The word 'fullerene' is derived from the word "Buckminsterfullerene," which refers specifically to the C_{60} molecule and is named after Buckminster Fuller, an architect who described and made famous the geodesic dome. C_{60} and C_{70} are the most common and easy to produce fullerenes.

Incidental nanosized materials: Nanomaterials that are the byproducts of human activity, such as combustion, welding, or grinding.

Intentionally produced nanomaterials: See **Engineered/manufactured nanoscale materials.**

Manufacturing processes: General term used to identify the variety of processes used in the production of the part. Processes may include plastic injection molding, vacuum forming, milling, stamping, casting, extruding, die-cutting, sewing, printing, packaging, polishing, grinding, metal spinning, welding, and so forth.

Nano-: a prefix meaning one billionth.

Nanobiology: A field of study combining biology and physics which looks at how nature works on the nanometer scale, particularly how transport takes place in biological systems. The interaction between the body and nanodevices are studied, for example, to develop processes for the body to regenerate bone, skin, and other damaged tissues.

Nanochemistry: A discipline focusing on the unique properties associated with the assembly of atoms or molecules on a nanometer scale. At this scale, new methods of carrying out chemical reactions are possible. Alternatively, it is the development of new tools, technologies and methodologies for doing chemistry in the nanolitre to femtolitre domains.

Nanoelectronics: Electronics on a nanometer scale, whether by current techniques or nanotechnology; includes both molecular electronics and nanoscale devices resembling today's semiconductor devices.

Nanomaterial: See **Engineered/manufactured nanoscale materials**

Nanometer: one billionth of a meter.

Nanoparticle: Free standing nanosized material, consisting of between tens to thousands of atoms.

Nanoscale: having dimensions measured in nanometers.

Nanoscience: the interdisciplinary field of science devoted to the advancement of nanotechnology.

Nanostructures: structures at the nanoscale; that is, structures of an intermediate size between molecular and microscopic (micrometer-sized) structures.

Nanotechnology: Research and technology development at the atomic, molecular or macromolecular levels, in the length scale of approximately 1 - 100 nanometer range; creating and using structures, devices and systems that have novel properties and functions because of their small and/or intermediate size; and the ability to control or manipulate on the atomic scale.

Nanotube: Tubular structure, carbon and non-carbon based, with dimensions in nanometer regime.

Nanowire: High aspect ratio structures with nanometer diameters that can be filled (nanorods) or hollow (nanotubes).

$PM_{0.1}$: Particulate matter less than 0.1 micrometers in diameter

$PM_{2.5}$: Particulate matter less than 2.5 micrometers in diameter

PM_{10}: Particulate Matter less than 10 micrometers in diameter

Quantum dot: A quantum dot is a closely packed semiconductor crystal comprised of hundreds or thousands of atoms, and whose size is on the order of a few nanometers to a few hundred nanometers. Changing the size of quantum dots changes their optical properties

Self-Assembled Monolayers on Mesoporous Supports (SAMMS): nanoporous ceramic materials that have been developed to remove contaminants from environmental media.

Self-assembly: The ability of objects to assemble themselves into an orderly structure. Routinely seen in living cells, this is a property that nanotechnology may extend to inanimate matter.

Self-replication: The ability of an entity such as a living cell to make a copy of itself.

Superlattice: nanomaterials composed of thin crystal layers. The properties (thickness, composition) of these layers repeat periodically.

Unintentionally produced nanomaterials: See Incidental nanosized materials

In: Nanotechnology and the Environment   ISBN: 978-1-60692-663-5
Editor: Robert V. Neumann   © 2010 Nova Science Publishers, Inc.

Appendix B

PRINCIPLES OF ENVIRONMENTAL STEWARDSHIP BEHAVIOR

U.S. Environmental Protection Agency

What does a good environmental steward do?
(based on statements by environmental stewards and others)

Exceeds required compliance. An environmental steward views environmental regulations only as a floor, not a target.

Protects natural systems and uses natural resources effectively and efficiently. An environmental steward considers and reduces the household, community, farm or company's entire environmental footprint. A steward safeguards and restores nature at home and elsewhere. A steward follows the pollution prevention hierarchy of acting first to prevent pollution at its source. A steward uses less toxic, more environmentally benign materials, uses local resources and conserves natural resources whenever possible. A steward reuses and recycles materials and wastes and seeks sustainability.

*Makes environment a key part of internal priorities, values and ethics, and leads by example.*Environmental stewards make decisions through their own volition that will prevent or minimize environmental harm. They anticipate, plan for, and take responsibility for economic, environmental and social consequences of actions. A steward approaches business strategies, policy planning, and life as an integrated dynamic with the environment. A steward acts in innovative ways, using all available tools and creating or adding value. A steward adopts holistic, systems approaches.

Holds oneself accountable. An environmental steward measures the effects of behavior on the environment and seeks progress. A steward applies an understanding of carrying capacity to measure progress and update objectives to achieve continuous improvement, often using indicators, environmental assessments, and environmental management systems.

Believes in shared responsibility. An environmental steward recognizes obligations and connections to all stakeholders- shareholders, customers, communities at home and elsewhere. For a company, this means being concerned with the full life cycle of products and services, beyond company boundaries, up and down the supply chain (including consumers and end-of-life). For a community, this means to protect the environment for all members and takes responsibility for effects on downstream air pollution, and effects of wastes disposed elsewhere. A steward operates with transparency. They encourage others to be collaborative stewards.

Invests in the future. An environmental steward anticipates the needs of future generations while serving the needs of the present generation. Their actions reflect possible changes in population, the economy and technology. A steward guides the development of technology to minimize negative environmental implications and maximize potential environmental stewardship applications. A steward values and protects natural and social capital. They seek preventative and long-term solutions in community development, business strategy, agricultural strategy, and household plans.

In: Nanotechnology and the Environment
Editor: Robert V. Neumann

ISBN: 978-1-60692-663-5
© 2010 Nova Science Publishers, Inc.

Appendix C

EPA's Nanotechnology Research Framework

U.S. Environmental Protection Agency

Nanotechnology has the potential to provide benefits to society and to improve the environment, both through direct applications to detect, prevent, and remove pollutants, the design of cleaner industrial processes and the creation of environmentally friendly products. However, some of the same unique properties that make manufactured nanoparticles beneficial also raise questions about the potential impacts of nanoparticles on human health and the environment.

Based on the fiscal year 2007 President's budget request of $8.6 million, EPA is developing a nanotechnology research strategy for fiscal years 2007-2012 that is problem-driven, focused on addressing the Agency's needs. The framework for this strategy, as outlined here, involves conducting research to understand whether nanoparticles, in particular those with the greatest potential to be released into the environment and/or trigger a hazard concern, pose significant risks to human health or ecosystems, considering the entire life cycle. EPA also will conduct research to identify approaches for detecting and measuring nanoparticles. This research framework is based on the recommendations from the EPA *Nanotechnology* White Paper and is consistent with the research needs identified by the Interagency Working Group on Nanotechnology Environmental and Health Implications, one of the working groups of the Nanoscale Science Engineering and Technology Subcommittee of the National Science and Technology Council.

While some studies have been done to determine potential toxicity of certain nanoparticles to humans and other organisms (both *in vivo* and *in vitro*), very little research has been performed on environmental fate and transport, transformation, and exposure potential. Research also is lacking on technologies and methods to detect and quantify nanomaterials in various environmental media. In addition, studies indicate that the toxicity of the nanomaterial will vary with size, surface charge, coating, state of agglomeration, etc. Therefore, in fiscal years 2007 and 2008, EPA will focus·on the following high priority areas: environmental fate, transport, transformation and exposure; and monitoring and detection methods. Resulting data will be used to inform and develop effects and exposure assessment methods and identify important points of releases for potential management. Specific activities will include:

- Identifying, adapting, and, where necessary, developing methods and techniques to measure nanomaterials from sources and in the environment
- Enhancing the understanding of the physical, chemical, and biological reactions nanomaterials undergo and the resulting transformations and persistence in air, soil and water
- Characterizing nanomaterials through their life cycle in the environment
- Providing the capability to predict significant exposure pathway scenarios
- Providing data to inform human health and ecological toxicity studies, as well as computational toxicological approaches, and aid in the development of the most relevant testing methods/protocols

Having laid a foundation for understanding possible material alterations under various conditions, EPA will direct a greater share of fiscal year 2009 and 2010 resources to exploring the effects, specifically toxicity of the altered materials as identified in the first two years. This approach will be informed and refined by case studies, initiating in fiscal year 2007, designed to elicit information on how EPA can address high-exposure-potential nanoparticles/nanomaterials. By 2011-2012, sufficient knowledge will result in the development of systematic and integrated approaches to assess, manage and communicate risks associated with engineered nanomaterials in the environment.

To complement its own research program, EPA is working with other federal agencies to develop research portfolios that address environmental and human health needs. In addition, the Agency is collaborating with academia and industry to fill knowledge gaps in these areas. Finally, the Agency is working internationally and is part of the Organization of Economic Cooperation and Development's efforts on the topic of the implications of manufactured nanomaterials.

In: Nanotechnology and the Environment
Editor: Robert V. Neumann

ISBN: 978-1-60692-663-5
© 2010 Nova Science Publishers, Inc.

Appendix D

EPA STAR Grants for Nanotechnology

U.S. Environmental Protection Agency

Through Science to Achieve Results (STAR) program in EPA's Office of Research and Development/National Center for Environmental Research, a number of nanotechnology research grants have been awarded. The table below shows nanotechnology grants funded though 2005. Additional grants focusing on implications of nanomaterials for the 2006 solicitation are in the process of final selection and funding by EPA, the National Science Foundation (NSF), the National Institute for Occupational Safety and Health (NIOSH), and the National Institute of Environmental Health Sciences (NIEHS). Information on funded grants, including abstracts and progress reports is available online at www.epa.gov/ncer/nano.

Grant #	Principal Investigator (PI)	Title	Institution	Year	Amount
RD829621	Bhattachar-yya, Dibakar	Membrane-Based Nanostructured Metals for Reductive Degradation of Hazardous Organic at Room Temperature	University of Kentucky	2002	$345,000
RD829606	Chen, Wilfred	Nanoscale Biopolymers with Tunable Properties for Improved Decontamination and Recycling of Heavy Metals	University of California, Riverside	2002	$390,000
RD829603	Chumanov, George	Plasmon Sensitized TiO_2 Nanoparticles as a Novel Photocatalyst for Solar Applications	Clemson University	2002	$320,000
RD829626	Diallo, Mamadou	Dendritic Nanoscale Chelating Agents: Synthesis, Characterization, Molecular Modeling and Environmental Applications	Howard University	2002	$400,000
RD829599	Gawley, Robert	Nanosensors for Detection of Aquatic Toxins	University of Arkansas	2002	$350,000
RD829622	Johnston, Murray	Elemental Composition of Freshly Nucleated Particles	University of Delaware	2002	$390,000

(Continued)

Grant #	Principal Investigator (PI)	Title	Institution	Year	Amount
RD829600	Larsen, Sarah	Development of Nanocrystalline Zeolite Materials as Environmental Catalysts: From Environmentally Benign Synthesis Emission Abatement	University of Iowa	2002	$350,000
RD829620	McMurry, Peter	Ion-Induced Nucleation of Atmospheric Aerosols	University of Minnesota	2002	$400,000
RD829624	Shah, S. Ismat	Synthesis, Characterization and Catalytic Studies of Transition Metal Carbide Nanoparticles as Environmental Nanocatalysts	University of Delaware	2002	$350,000
RD829604	Shih, Wan	Ultrasensitive Pathogen Quantification in Drinking Water Using Highly Piezoelectric PMN-PT Microcantilevers	Drexel University	2002	$400,000
RD829602	Sigmund, Wolfgang	Simultaneous Environmental Monitoring and Purification through Smart Particles	University of Florida	2002	$390,000
RD829601	Strongin, Daniel	A Bioengineering Approach to Nanoparticle Based Environmental Remediation	Temple University	2002	$399,979
RD829623	Tao, Nongjian	A Nanocontact Sensor for Heavy Metal Ion Detection	Arizona State University	2002	$375,000
RD829619	Trogler, William	Nanostructured Porous Silicon and Luminescent Polysiloles as Chemical Sensors for Carcinogenic Chromium (VI) and Arsenic (V)	University of California, San Diego	2002	$400,000
RD829605	Velegol, Darrell	Green Engineering of Dispersed Nanoparticles: Measuring and Modeling Nanoparticles Forces	Pennsylvania State University	2002	$370,000
RD829625	Zhang, Wei-xian	Nanoscale Bimetallic Particles for In Situ Remediation	Lehigh University	2002	$300,000
RD830907	Anderson, Anne	Metal Biosensors: Development and Environmental Testing	Utah State University	2003	$336,000
RD830910	Beaver, Earl	Implications of Nanomaterials Manufacture and Use: Development of a Methodology for Screening Sustainability	BRIDGES to Sustainability	2003	$99,741
RD830904	Drzal, Lawrence	Sustainable Biodegradable Green Nanocomposites from Bacterial Bioplastic for Automotive Applications	Michigan State University	2003	$369,000
RD830902	Kan, Edwin	Neuromorphic Approach to Molecular Sensing with Chemoreceptive Neuron MOS Transistors (CnMOS)	Cornell University	2003	$354,000
RD830909	Kilduff, James	Graft Polymerization as a Route to Control Nanofiltration Membrane Surface Proper- ties to Manage Risk of EPA Candidate Contaminants and Reduce NOM Fouling	Rensselaer Polytechnic Institute	2003	$349,000
RD830905	Lave, Lester	Environmental Implications of Nanotechnology	Carnegie Mellon University	2003	$100,000
RD830911	Lavine, Barry	Compound Specific Imprinted Nanospheres for Optical Sensing	Oklahoma State University	2003	$323,000
RD830898	Lowry, Gregory	Functional Fe(0)-Based Nanoparticles for In Situ Degradation of DNAPL Chlorinated Organic Solvents	Carnegie Mellon University	2003	$358,000

(Continued)

Grant #	Principal Investigator (PI)	Title	Institution	Year	Amount
RD830908	Masten, Susan	Use of Ozonation in Combination with Nanocomposite Ceramic Membranes for Controlling Disinfection By-Products	Michigan State University	2003	$353,959
RD830901	Mitra, Somenath	Micro-Integrated Sensing System (μ-ISS) by Controlled Assembly of Carbon Nanotubes on MEMS Structures	New Jersey Institute of Technology	2003	$346,000
RD830903	Sabatini, David	Nanostructured Microemulsions as Alternative Solvents to VOCs in Cleaning Technologies and Vegetable Oil Extraction	University of Oklahoma, Norman	2003	$315,000
RD830906	Sadik, Omowunmi	Advanced Nanosensors for Continuous Monitoring of Heavy Metals	State University of New York, Binghamton	2003	$351,000
RD830896	Senkan, Selim	Nanostructured Catalytic Materials for NOx Reduction Using Combinatorial Methodologies	University of California, Los Angeles	2003	$356,000
RD830899	Subramanian, Vivek	Low Cost Organic Gas Sensors on Plastic for Distributed Environmental Monitoring	University of California, Berkeley	2003	$328,000
RD830900	Wang, Joseph	Nanomaterial-Based Microchip Assays for Continuous Environmental Monitoring	Arizona State University	2003	$341,000
RD830897	Winter, William	Ecocomposites Reinforced with Cellulose Nanoparticles: An Alternative to Existing Petroleum-Based Polymer Composites	State University of New York, Syracuse	2003	$320,000
RD831722	Elder, Alison	Iron Oxide Nanoparticle-Induced Oxidative Stress and Inflammation	University of Rochester	2004	$350,000
RD831716	Ferguson, P. Lee	Chemical and Biological Behavior of Carbon Nanotubes in Estuarine Sedimentary Systems	University of South Carolina	2004	$349,740
RD831717	Grassian, Vicki	A Focus on Nanoparticulate Aerosol and Atmospherically Processed Nanoparticulate Aerosol	University of Iowa	2004	$350,000
RD831712	Holden, Patricia	Transformations of Biologically Conjugated CdSe Quantum Dots Released Into Water and Biofilms	University of California, Santa Barbara	2004	$343,853
RD831721	Huang, Chin-pao	Short-Term Chronic Toxicity of Photocatalytic Nanoparticles to Bacteria, Algae, and Zooplankton	University of Delaware	2004	$349,876
RD831719	Hurt, Robert	Physical and Chemical Determinants of Nanofiber/Nanotube Toxicity	Brown University	2004	$350,000
RD831715	Monteiro-Riviere, Nancy	Evaluated Nanoparticle Interactions with Skin	North Carolina State University	2004	$340,596
RD831714	Pinkerton, Kent	Health Effects of Inhaled Nanomaterials	University of California, Davis	2004	$349,998
RD831718	Tomson, Mason	Absorption and Release of Contaminants onto Engineered Nanoparticles	Rice University	2004	$348,747
RD832531	Turco, Ronald	Repercussion of Carbon Based Manufactured Nanoparticles on Microbial Processes in Environmental Systems	Purdue University	2004	$350,000

(Continued)

Grant #	Principal Investigator (PI)	Title	Institution	Year	Amount
RD831723	Veranth, John	Responses of Lung Cells to Metals in Manufactured Nanoparticles	University of Utah	2004	$344,748
RD831713	Westerhoff, Paul	The Fate, Transport, Transformation and Toxicity of Manufactured Nanomaterials in Drinking Water	Arizona State University	2004	$349,881
GR8322-25\	Zhang, Wei-xian	Transformation of Halogenated PBTs with Nanoscale Bimetallic Particles	Lehigh University	2004	$325,000
RD832534	Alvarez, Pedro	Microbial Impacts of Engineered Nanoparticles	William Marsh Rice University	2005	$375,000
RD832531	Asgharian, Bahman	Mechanistic Dosimetry Models of Nanomaterial Deposition in the Respiratory Tract	CIIT Centers for Health Research	2005	$375,000
RD832532	Bakshi, Bhavik	Evaluating the Impacts of Nanomanufacturing via Thermodynamic and Life Cycle Analysis	Ohio State University	2005	$375,000
	Barber, David	Uptake and Toxicity of Metallic Nanoparticles in Freshwater Fish	University of Florida	2005	NSF
RD832530	Bertsch, Paul	The Bioavailability, Toxicity, and Trophic Transfer of Manufactured ZnO2 Nanoparticles: A View from the Bottom	University of Georgia	2005	$363,680
RD832635	Bonzongo, Jean-Claude	Assessment of the Environmental Impacts of Nanotechnology on Organisms and Ecosystems	University of Florida	2005	$375,000
RD832536	Colvin, Vicki	Structure-Function Relationships in Engineered Nanomaterial Toxicity	William Marsh Rice Unibersity	2005	$375,000
	Cunningham, Mary Jane	Gene Expression Profiling of Single-Walled Carbon Nanotubes: A Unique Safety Assessment Approach	Houston Advanced Research Center	2005	NSF
RD832525	Diallo, Mamadou	Cellular Uptake and Toxicity of Dendritic Nanomaterials: An Integrated Physicochemical and Toxicogenomics Study	California Institute of Technology	2005	$375,000
GR832382	Gawley, Robert	Nanosensors for Detection of Saxitoxin	University of Arkansas	2005	$340,000
RD832528	Gordon, Terry	Role of Particle Agglomeration in Nanoparticle Toxicity	New York University School of Medicine	2005	$375,000
GR832371	Heiden, Patricia	A Novel Approach to Prevent Biocide Leaching	Michigan Technological University	2005	$333,130
RD832529	Kibbey, Tohren	Hysteretic Accumulation and Release of Nanomaterials in the Vadose Zone	University of Oklahoma	2005	$375,000
RD832526	Kim, Jaehong	Fate and Transformation of C60 Nanoparticles in Water Treatment Processes	Georgia Institute of Technology	2005	$375,000
GR832372	Kit, Kevin	Nanostructured Membranes for Filtration, Disinfection and Remediation of Aqueous and Gaseous Systems	University of Tennessee	2005	$349,200
GR832374	Lu, Yunfeng	Novel Nanostructured Catalysts for Environmental Remediation of Chlorinated Compounds	Tulane University	2005	$320,000

(Continued)

Grant #	Principal Investigator (PI)	Title	Institution	Year	Amount
	Marr, Linsey	Cross-Media Environmental Transport, Transformation, and Fate of Manufactured Carbonaceous Nanomaterials	Virginia Polytechnic Institute and State University	2005	NSF
RD832527	McDonald, Jacob	Chemical Fate, Biopersistence, and Toxicology of Inhaled Metal Oxide Nanoscale Materials	Lovelace Biomedical & Environmental Research Institute	2005	$375,000
GR832375	Mulchandani, Ashok	Conducting-Polymer Nanowire Immunosensor Arrays for Microbial Pathogens	University of California, Riverside	2005	$320,000
R01OH8806	O'Shaughnessy, Patrick	Assessment Methods for Nanoparticles in the Workplace	University of Iowa	2005	NIOSH
RD832535	Pennell, Kurt	Fate and Transport of C60 Nanomaterials in Unsaturated and Saturated Soils	Georgia Institute of Technology	2005	$375,000
RD832537	Perrotta, Peter	Effects of Nanomaterials on Human Blood Coagulation	West Virginia University	2005	$375,000
RD832533	Theodorakis, Chris	Acute and Developmental Toxicity of Metal Oxide Nanoparticles to Fish and Frogs	Southern Illinois University	2005	$375,000
R01OH8-807	Xiong, Judy	Monitoring and Characterizing Airborne Carbon Nanotube Particles	New York University School of Medicine	2005	NIOSH
GR832373	Zhao, Dongye	Synthesis and Application of a New Class of Stabilized Nanoscale Iron Particles for Rapid Destruction of Chlorinated Hydrocarbons in Soil and Groundwater	Auburn University	2005	$280,215
				Total EPA	$22,613,343
		70 TOTAL - 65 STAR, 3 NSF, 2 NIOSH			

In: Nanotechnology and the Environment

ISBN: 978-1-60692-663-5

Editor: Robert V. Neumann

© 2010 Nova Science Publishers, Inc.

Appendix E

LIST OF NANOTECHNOLOGY WHITE PAPER EXTERNAL PEER REVIEWERS AND THEIR AFFILIATIONS

U.S. Environmental Protection Agency

Pratim Biswas, Ph.D.
Departments of Chemical and Civil Engineering
Environmental Engineering Science Program
Washington University in St. Louis

Richard A. Denison, Ph.D.
Senior Scientist
Environmental Defense

Rebecca D. Klaper, Ph.D.
Great Lakes WATER Institute
University of Wisconsin, Milwaukee

Igor Linkov, Ph.D.
Senior Scientist
Cambridge Environmental Inc
Current Affiliation: Managing Scientist, INTERTOX, Inc.

Andrew D. Maynard, Ph.D.
Chief Science Advisor
Project on Emerging Nanotechnologies
Woodrow Wilson International Center for Scholars

Vladamir V. Murashov, Ph.D.
Special Assistant to the Director

National Institute for Occupational Safety and Health

Stephen S. Olin, Ph.D.
Deputy Director
International Life Sciences Institute (ILSI) Research Foundation

Jennifer B. Sass, Ph.D.
Senior Scientist, Health and Environment
Natural Resources Defense Council

Donald A. Tomalia, Ph.D.
President & Chief Technical Officer
Dendritic Nanotechnologies, Inc.

Nigel J. Walker, Ph.D.
National Institute of Environmental Health Sciences
National Institutes of Health
David B. Warheit, Ph.D
Senior Research Toxicologist, Inhalation Toxicology
E.I. du Pont de Nemours & Co., Inc.
Haskell Laboratory

INDEX

D

M

N

O

P

T

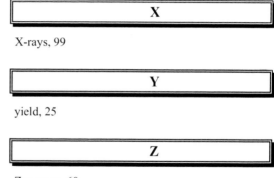